Fracture Mechanics Applications

Edited by Hayri Baytan Ozmen and H. Ersen Balcioglu

Published in London, United Kingdom

IntechOpen

Supporting open minds since 2005

Fracture Mechanics Applications
http://dx.doi.org/10.5772/intechopen.77417
Edited by Hayri Baytan Ozmen and H. Ersen Balcioglu

Contributors
Ricardo Castedo, Maria Chiquito, Anastasio P. Santos, Lina M. Lopez, Namas Chandra, Eren Alay, Molly Townsend, Maciej Skotak, Hayri Baytan Ozmen, H. Ersen Balcioglu, Tian-You Fan, Hui Cheng, Francisco Casanova-Del-Angel, Daniel Kimpfbeck, Zoltan Major, Matei Miron, Hiroshi Noguchi, Tatsujiro Miyazaki, Shigeru Hamada, Chatarina Niken, Daniel Hernández-Galicia, Xochicale-Rojas Hugo Alberto

First published in London, United Kingdom, 2020 by IntechOpen
IntechOpen is the global imprint of INTECHOPEN LIMITED, registered in England and Wales, registration number: 11086078, 5 Princes Gate Court, London, SW7 2QJ, United Kingdom
Printed in Croatia

British Library Cataloguing-in-Publication Data
A catalogue record for this book is available from the British Library

Additional hard and PDF copies can be obtained from orders@intechopen.com

Fracture Mechanics Applications
Edited by Hayri Baytan Ozmen and H. Ersen Balcioglu
p. cm.
Print ISBN 978-1-83880-448-0
Online ISBN 978-1-83880-449-7
eBook (PDF) ISBN 978-1-83968-771-6

Meet the editors

Dr. Hayri Baytan Ozmen is currently an associate professor in the Department of Civil Engineering, Usak University, Turkey. He graduated from the Civil Engineering Department of the Middle East Technical University, Turkey, in 2001. He received his PhD in the same field from Pamukkale University in 2011. His research interests includes reinforced concrete structures, earthquake engineering, seismic evaluation, and retrofit. He has more than sixty-five research papers published in international journals and conferences and has conducted and been involved in more than ten national and international research projects. He performed seismic evaluation or design of seismic retrofit systems for more than 150 RC buildings and provided consultancy for structural engineering studies. He is the editor in chief of an international journal on materials and structural engineering.

Dr. H. Ersen Balcioglu is a research assistant in the Mechanical Engineering Department at Usak University, Turkey. He completed an MSc in 2009 at Celal Bayar University. He received his PhD in Mechanical Engineering from Usak University in 2017. His research interests are applications in fiber-reinforced composites (FRCs), mechanical properties of FRCs, finite element analysis, fracture mechanics, and fatigue behavior of composites. He has published more than thirty-five peer-reviewed publications (cited more than 100 times) and more than ten international symposiums. He also worked as a researcher in four scientific projects. He is the editor of an international journal on materials and structural engineering.

Contents

Preface

Fracture mechanics deals with the cracking behavior of materials, and cracking defines the limit state for many components of engineering systems. Therefore, fracture mechanics is a field that provides useful tools for many engineering practitioners and designers from various disciplines. It is a relatively new branch of mechanics and more advanced in theoretical basis rather than practical applications. However, due to its useful approach in handling failure of materials, its use is expected to be more widespread in the near future.

Fracture mechanics principles can help us design more robust components ensuring safer airplanes, space shuttles, ships, cranes, buildings, bridges and other mechanical systems for use in medicine, for example, destroying arterial plaques or kidney stones. In short, application of fracture mechanics has potential to give us tools in solution of many design and analysis problems. In this context, this book introduces and supports the understanding of fracture mechanics principles to enhance the knowledge and application of specialists in all related areas of engineering and science.

Fracture Mechanics Applications contains eight chapters written by researchers and experts in the field. It is designed for graduate students, researchers, and practicing engineers.

Chapter 1 introduces rupture theory of soft-matter quasicrystals. Chapters 2 provides insight into application of the J-integral for the fracture assessment of welded polymeric components. Chapter 3 describes fatigue limit reliability analysis for notched material with some kinds of dense inhomogeneities using fracture mechanics. Chapter 4 deals with the fracture behavior of pure and hybrid intraply knitted fabric-reinforced polymer composites. Chapter 5 explains the construction of an artificial neural network to detect structural damage. Chapter 6 introduces the mechanism of misalignment of saw cutting cracks in concrete pavement. Chapter 7 presents the effect of geometrical factors on the surface pressure distribution on a human phantom model following shock exposure. Finally, Chapter 8 discusses blast effects on different structural elements.

We would like to express our deepest appreciation to academicians, scientists, and our colleagues and friends globally who have significantly contributed in the broad field of fracture mechanics, and especially to the authors of the chapters in this book. We are also grateful to IntechOpen for publishing the book, which we hope to be a noteworthy contribution to the current literature.

Dr. Hayri Baytan Özmen and Dr. H. Ersen Balcioglu
Usak University,
Uşak, Turkey

Chapter 1

Probe on Rupture Theory of Soft-Matter Quasicrystals

Hui Cheng and Tian-You Fan

Abstract

In this chapter, a probe on an important aspect, rupture theory of soft matter, is discussed, in which the soft matter and soft-matter quasicrystals are introduced at first. Then, we discuss the behaviour of the matter. For the soft-matter quasicrystals, there are basic equations describing their dynamics; due to the work of the second author of this chapter, this provides a fundamental for studying the rupture feature quantitatively. For general soft matter, there are no such equations so far, whose rupture theory quantitatively is not available at moment. The discussion on the soft-matter quasicrystals may provide a reference for other branches of soft matter.

Keywords: soft matter, quasicrystals, generalized dynamics, equation of state, crack, stress intensity factor, generalized Griffith criterion of rupture

1. Introduction

Soft-matter quasicrystals belong to a category of soft matter. It is well known that the soft matter is an intermediate phase between solid and liquid, which presents the behaviour of both solid and liquid, the first nature of which is fluidity, and behaves as a complex structure. Hence, the soft matter is named as a complex liquid or a structured liquid. Soft-matter quasicrystals are soft-matter with quasi-periodic symmetry. In this sense, they are a category of soft matter with highest symmetry so far. The high symmetry of the matter presents symmetry breaking and leads to importance of elementary excitations. This helps us to set up their dynamic equations and provide the fundamental for the rupture study.

2. Soft matter and the rupture problem

The liquid crystals, polymers, colloids, surfactants, and so on are in common named as soft matter. The 12-fold symmetry quasicrystals were observed most frequently in the soft matter [1–7]. At the same time, cracks in soft matter [8, 9] cannot be ignored and should be prevented [10] from the material safety. Other crack and rupture problems in soft matter can be referred from [11–18]. This shows that the study on crack and rupture for soft matter including soft-matter quasicrystals is very important.

It is well known that the failure of brittle structural materials has been well studied. According to the work of Griffith [19], the existence and propagation of crack is the reason of the failure of these materials. Griffith used the exact solution of a crack in an infinite plate and calculated the crack strain energy. The energy is the function of crack size. He further calculated crack energy release rate and suggested that when the release rate equals to the surface tensile of the material, the solid will be a failure. His hypothesis was proved by experiments. This is the famous Griffith criterion. Afterward, the classical Griffith theory was developed by Irwin et al. for studying quasi-brittle failure, where the strain energy release rate was replaced by so-called stress intensity factor and the surface tensile was replaced by the fracture toughness of the material. This is the engineering approach of the Griffith theory, and played an important role in engineering application. The failure of ductile materials is also related with the existence and propagation of crack, but the problem has not been well studied due to the plastic deformation around the crack tip. The plastic deformation is a nonlinear irreducible process. The problem is extremely complex physically and mathematically. The failure of soft matter will be more complex than the ductile structural materials because of the existence of the fluidity. Especially, the experimental results are few of reported. As a most preliminary probe for studying the problem of soft matter, we try to draw from the idea of the classical Griffith work, i.e., to study the crack stability/instability, we can use the so-called following Griffith-Irwin criterion

$$K_1(a,\sigma) = K_{IC}(T,f(s)) \tag{1}$$

in which $K_1(a,\sigma)$ represents elastic stress intensity factor, which a function of crack size a and applied external stress σ, can be determined by stress analysis of cracked materials, and the $K_{IC}(T,f(s))$ fracture toughness, a material constant but influenced by temperature and the structure factor $f(s)$, and the suffice I expresses mode I, i.e., the opening mode fracture (and the mode II is shearing mode, or slip mode fracture, and mode III is longitudinal shear mode, or tearing mode fracture, we here discuss only the opening mode). If the value of $K_1(a,\sigma)$ is greater than that of $K_{IC}(T,f(s))$, then the crack will propagate and the material will fracture.

Of course, the criterion (1) is only a reference for the soft matter, and a further analysis will be given in the subsequent sections.

3. Soft-matter quasicrystals and their generalized dynamics

There are a quite lot of references concerning the crack and rupture problems in soft matter; however, the quantitative analyses are not so much, because most branches of soft matter science are in qualitative stage so far. Either theoretical research or engineering application, the rupture problem of the soft matter needs a quantitative analysis.

Recently, generalized dynamics of soft-matter quasicrystals has been developed [20–24], which may become another quantitative branch in soft matter apart from the liquid crystals science. The generalized dynamics of soft-matter quasicrystals provides a tool for analyzing quantitatively rupture problem of the matter, whose result may be references of other categories of soft matter.

Soft-matter quasicrystals look like other categories of soft matter which belong to complex fluid; at mean time they present highly symmetry. For the high ordered phase, the symmetry breaking and elementary excitation principle are important. By using the Landau-Anderson [25, 26] symmetry breaking and elementary excitation principle, there are phonon and phason elementary excitations. As a class of soft matter, the fluidity is the substantive nature of the soft-matter quasicrystals; so Fan [20–24] introduced another elementary excitation—fluid phonon apart from phonons and phasons; of course, the concept of the fluid phonon is originated from the Landau School [27]. The introducing of fluid phonon requires a supplemented equation and the equation of state as well, which is also completed by Fan and co-worker [28]. With these bases, the generalized dynamics of soft-matter quasicrystals is set up.

For the need of the present chapter, we list the two-dimensional equations of the dynamics of soft-matter quasicrystals of 5- and 10-fold symmetry as follows:

$$\left.\begin{array}{l}
\dfrac{\partial \rho}{\partial t}+\nabla\cdot(\rho\mathbf{V})=0 \\
\dfrac{\partial(\rho V_x)}{\partial t}+\dfrac{\partial(V_x\rho V_x)}{\partial x}+\dfrac{\partial(V_y\rho V_x)}{\partial y}=-\dfrac{\partial p}{\partial x}+\eta\nabla^2 V_x+\dfrac{1}{3}\eta\dfrac{\partial}{\partial x}\nabla\cdot\mathbf{V} \\
+M\nabla^2 u_x+(L+M-B)\dfrac{\partial}{\partial x}\nabla\cdot\mathbf{u} \\
+R_1\left(\dfrac{\partial^2 w_x}{\partial x^2}+2\dfrac{\partial^2 w_y}{\partial x\partial y}-\dfrac{\partial^2 w_x}{\partial y^2}\right)-R_2\left(\dfrac{\partial^2 w_y}{\partial x^2}-2\dfrac{\partial^2 w_x}{\partial x\partial y}-\dfrac{\partial^2 w_y}{\partial y^2}\right) \\
-(\mathrm{A}-\mathrm{B})\dfrac{1}{\rho_0}\dfrac{\partial\delta\rho}{\partial x} \\
\dfrac{\partial(\rho V_y)}{\partial t}+\dfrac{\partial(V_x\rho V_y)}{\partial x}+\dfrac{\partial(V_y\rho V_y)}{\partial y}=-\dfrac{\partial p}{\partial y}+\eta\nabla^2 V_y+\dfrac{1}{3}\eta\dfrac{\partial}{\partial y}\nabla\cdot\mathbf{V} \\
+M\nabla^2 u_y+(L+M-B)\dfrac{\partial}{\partial y}\nabla\cdot\mathbf{u} \\
+R_1\left(\dfrac{\partial^2 w_y}{\partial x^2}-2\dfrac{\partial^2 w_x}{\partial x\partial y}-\dfrac{\partial^2 w_y}{\partial y^2}\right)+R_2\left(\dfrac{\partial^2 w_x}{\partial x^2}+2\dfrac{\partial^2 w_y}{\partial x\partial y}-\dfrac{\partial^2 w_x}{\partial y^2}\right) \\
-(\mathrm{A}-\mathrm{B})\dfrac{1}{\rho_0}\dfrac{\partial\delta\rho}{\partial y} \\
\dfrac{\partial u_x}{\partial t}+V_x\dfrac{\partial u_x}{\partial x}+V_y\dfrac{\partial u_x}{\partial y}=V_x+\Gamma_{\mathbf{u}}[M\nabla^2 u_x+(L+M)\dfrac{\partial}{\partial x}\nabla\cdot\mathbf{u}+ \\
R_1\left(\dfrac{\partial^2 w_x}{\partial x^2}+2\dfrac{\partial^2 w_y}{\partial x\partial y}-\dfrac{\partial w_x}{\partial y^2}\right)-R_2\left(\dfrac{\partial^2 w_y}{\partial x^2}-2\dfrac{\partial^2 w_x}{\partial x\partial y}-\dfrac{\partial^2 w_y}{\partial y^2}\right)\Bigg] \\
\dfrac{\partial u_y}{\partial t}+V_x\dfrac{\partial u_y}{\partial x}+V_y\dfrac{\partial u_y}{\partial y}=V_y+\Gamma_{\mathbf{u}}[M\nabla^2 u_y+(L+M)\dfrac{\partial}{\partial y}\nabla\cdot\mathbf{u}+ \\
R_1\left(\dfrac{\partial^2 w_y}{\partial x^2}-2\dfrac{\partial^2 w_x}{\partial x\partial y}-\dfrac{\partial^2 w_y}{\partial y^2}\right)+R_2\left(\dfrac{\partial^2 w_x}{\partial x^2}+2\dfrac{\partial^2 w_y}{\partial x\partial y}-\dfrac{\partial^2 w_x}{\partial y^2}\right)\Bigg] \\
\dfrac{\partial w_x}{\partial t}+V_x\dfrac{\partial w_x}{\partial x}+V_y\dfrac{\partial w_x}{\partial y}=\Gamma_{\mathbf{w}}[K_1\nabla^2 w_x+ \\
R_1\left(\dfrac{\partial^2 u_x}{\partial x^2}-2\dfrac{\partial^2 u_y}{\partial x\partial y}-\dfrac{\partial^2 u_x}{\partial y^2}\right)+R_2\left(\dfrac{\partial^2 u_y}{\partial x^2}+2\dfrac{\partial^2 u_x}{\partial x\partial y}-\dfrac{\partial^2 u_y}{\partial y^2}\right)\Bigg] \\
\dfrac{\partial w_y}{\partial t}+V_x\dfrac{\partial w_y}{\partial x}+V_y\dfrac{\partial w_y}{\partial y}=\Gamma_{\mathbf{w}}[K_1\nabla^2 w_y+ \\
R_1\left(\dfrac{\partial^2 u_y}{\partial x^2}+2\dfrac{\partial^2 u_x}{\partial x\partial y}-\dfrac{\partial^2 u_y}{\partial y^2}\right)-R_2\left(\dfrac{\partial^2 u_x}{\partial x^2}-2\dfrac{\partial^2 u_y}{\partial x\partial y}-\dfrac{\partial^2 u_x}{\partial y^2}\right)\Bigg] \\
p=3\dfrac{k_B T}{l^3}\left(\dfrac{\rho}{\rho_0}+\dfrac{\rho^2}{{\rho_0}^2}+\dfrac{\rho^3}{{\rho_0}^3}\right)
\end{array}\right\} \qquad (2)$$

in which u_i denotes the phonon field, w_i the phason field, V_i the fluid velocity field, C_{ijkl} is the phonon elastic constant tensor, K_{ijkl} phason elastic constant tensor, and R_{ijkl} and R_{klij} are the phonon-phason coupling elastic constant tensors.

A and B are the constants describing density variation, $g = \rho\mathbf{V}$, and η is the fluid viscosity, k_B is the Boltzmann constant, T is the absolute temperature, and $l \sim 10 - 100\,\text{nm}$ is the characteristic size of the soft matter, respectively.

4. An example of stress analysis of soft-matter quasicrystals

With Eq. (2), we can carry out a stress analysis of some fundamental specimens with crack of soft-matter quasicrystals; we here give only a computational example as shown in **Figure 1**.

If we want to obtain further information on deformation and motion of the material, we must solve the equations under appropriate initial and boundary conditions. To solve the problem, a specimen made by the matter should be optioned, which is subjected to some initial and boundary conditions. Here, the corresponding conditions of the specimen shown in **Figure 1** are as follows:

$$\begin{aligned}
&\mathrm{t} = 0 : V_x = V_y = 0, \mathrm{u}_x = \mathrm{u}_y = 0, \mathrm{w}_x = w_y = 0, p = p_0;\\
&y = \pm H, |x| < W : V_x = V_y = 0, \sigma_{yy} = \sigma_0 f(t), \sigma_{yx} = 0, H_{yy} = H_{yx} = 0, p = p_0;\\
&x = \pm W, |y| < H : V_x = V_y = 0, \sigma_{xx} = \sigma_{xy} = 0, H_{xx} = H_{xy} = 0, p = p_0;\\
&y = 0, |x| < a : V_x = V_y = 0, \sigma_{yy} = \sigma_{yx} = 0, H_{yy} = H_{yx} = 0, p = p_0
\end{aligned} \tag{3}$$

In the present computation we take, $2H = 0.01\,\text{m}$, $2W = 0.01\,\text{m}$, $2a = 0.0024\,\text{m}$, $\sigma_0 = 0.01\,\text{MPa}$, $\rho_0 = 1.5 \times 10^3\,\text{kg/m}^3$, $\eta = 0.1\,\text{Pa}\cdot\text{s}\,(\text{or } 0.2\,\text{Pa}\cdot\text{s})$, $\zeta = 0$, $L = 10\,\text{MPa}$, $M = 4\,\text{MPa}$, $K_1 = 0.5\text{L}$, $R = R_1 = 0.04\,\text{M}\,(\text{or } 0.06\text{M})$, $R_2 = 0$, $\Gamma_u = 4.8 \times 10^{-17}\,\text{m}^3\cdot\text{s/kg}$, $\Gamma_w = 4.8 \times 10^{-19}\,\text{m}^3\cdot\text{s/kg}$, $\text{A} \sim 0.2\,\text{MPa}$, $\text{B} \sim 0.2\,\text{MPa}$, and p_0 denotes 1 atm.

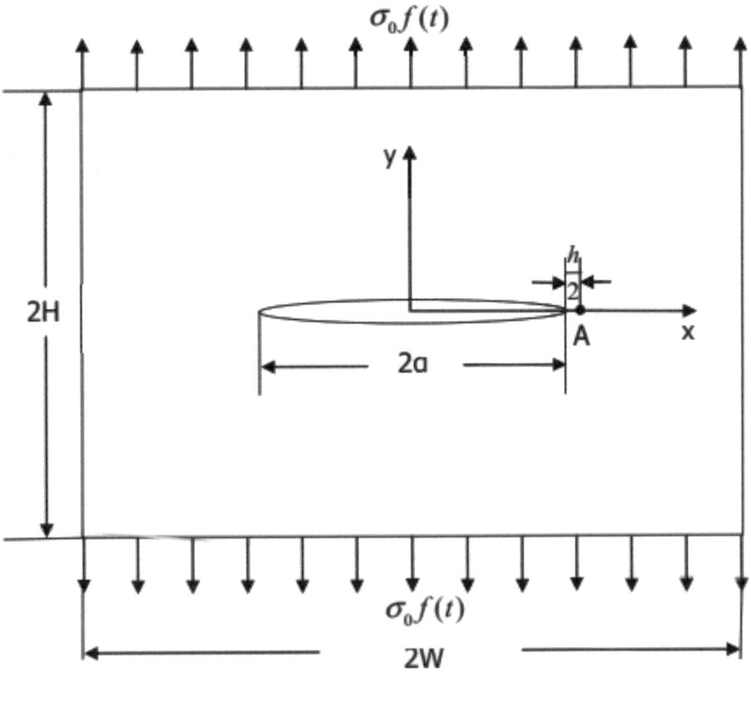

Figure 1.
Specimen of soft-matter quasicrystals of 5- and 10-fold symmetries with a Griffith crack under tension.

The initial and boundary value problem (3) of Eq. (2) can be solved by the finite difference method to solve the boundary value problem (**Figure 2**), e.g.,

$$\frac{\partial^2 u_x}{\partial x^2} = \frac{u_x(i+1,j) - 2u_x(i,j) + u_x(i-1,j)}{h^2},$$
$$\frac{\partial^2 u_x}{\partial y^2} = \frac{u_x(i,j+1) - 2u_x(i,j) + u_x(i,j-1)}{h^2},$$
$$\frac{\partial u_x}{\partial t} = \frac{u_x(k+1) - u_x(k)}{\tau}.$$

so determine the phonon and phason displacement fields and fluid phonon velocity filed, then the phonon and phason strain tensors.

$$\varepsilon_{ij} = \frac{1}{2}\left(\frac{\partial u_i}{\partial x_j} + \frac{\partial u_j}{\partial x_i}\right), w_{ij} = \frac{\partial w_i}{\partial x_j}, \tag{4}$$

and the fluid phonon deformation rate tensor

$$\dot{\xi}_{ij} = \frac{1}{2}\left(\frac{\partial V_i}{\partial x_j} + \frac{\partial V_j}{\partial x_i}\right) \tag{5}$$

and according to constitutive laws

$$\left.\begin{aligned} \sigma_{ij} &= C_{ijkl}\varepsilon_{ik} + R_{ijkl}w_{kl}, \\ H_{ij} &= K_{ijkl}w_{ij} + R_{klij}\varepsilon_{kl}, \\ p_{ij} &= -p\delta_{ij} + \sigma'_{ij} = -p\delta_{ij} + 2\eta\dot{\xi}_{kl}, \end{aligned}\right\} \tag{6}$$

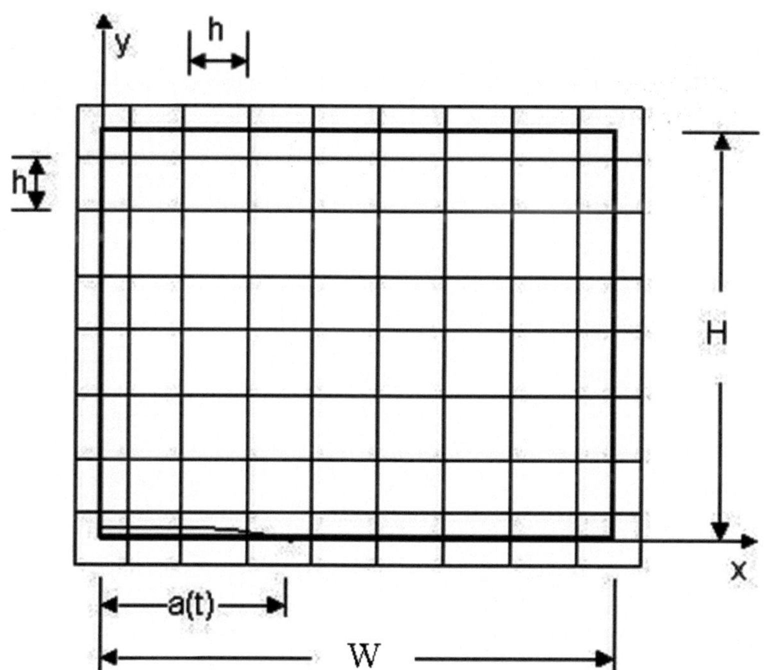

Figure 2.
The scheme of grid of the difference for a part of the specimen.

we obtain the phonon stresses σ_{ij} and fluid phonon stresses p_{ij}, respectively, in which recall C_{ijkl} the phonon elastic constant tensor, K_{ijkl} phason elastic constant tensor, R_{ijkl} and R_{klij} the phonon-phason coupling elastic constant tensors, etc., refer to [24].

Due to the complexity of the equations, the computation is complex too. For the dynamic problems, i.e., there are manmade damping terms $\theta\rho$, $\theta\rho V_x$, and so on, the iterative computation is easily stable, and for the static problems, we still take the iterative computation; however, the choosing of the manmade damping coefficient θ depends upon experience.

The computational results are mostly dependent on the ratio value of σ_0/p_0, i.e., the ratio of amplitudes of the phonon stress and fluid phonon stress apart from other factors.

5. Significance of fluid stress to crack initiation of growth and crack propagation

In the crack and fracture of brittle or quasibrittle failure of structural materials, the stress analysis is a basic task, from which one can determine the stress intensity factor, and then use the Griffith-Irwin criterion to analyze the crack stability/instability. For the Mode I crack, the tensile stress $\sigma_{yy}(\mathrm{x}, \mathrm{y}, \mathrm{t})$ is the most important, which leads to the crack surface opening and so the crack propagation.

After our computation, in soft matter including soft-matter quasicrystals, apart from the phonon stress $\sigma_{yy}(\mathrm{x}, \mathrm{y}, \mathrm{t})$, there is the fluid phonon stress $p_{yy}(\mathrm{x}, \mathrm{y}, \mathrm{t})$ which is pressure and leads to crack closing (**Figures 3** and **4**).

In principle, the Griffith theory holds for describing brittle and quasi-brittle rupture of structural materials (or engineering materials). However, for soft matter including soft-matter quasicrystals, there is fundamental difference with the structural materials. The key reason about this is the existence of fluid effect; in terminology of soft-matter quasicrystal study, it is also called the existence of fluid phonon. According to our analysis, the effect of fluid stress intensity factor is opposite to the elastic stress intensity factor (or by using the terminology of soft-matter quasicrystals study, the elastic stress intensity factor is also called the phonon stress

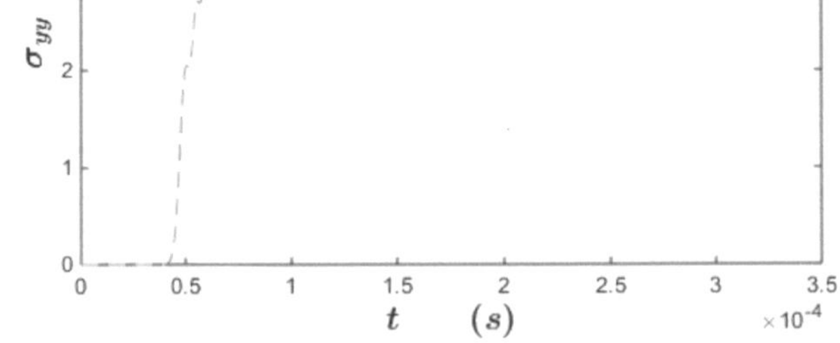

Figure 3.
Normal stress $\sigma_{yy}(x, 0, t)$ of phonon field at the point A of specimen versus time.

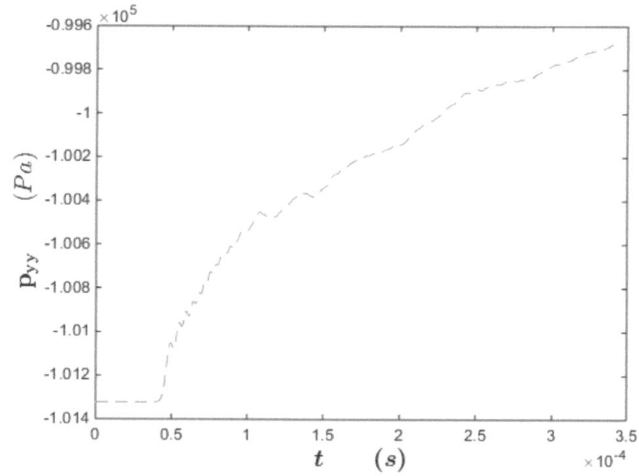

Figure 4.
Normal stress $p_{yy}(x, 0, t)$ of fluid field at the point A of specimen versus time.

intensity factor). K_1^{phonon} denotes the phonon stress intensity factor and K_1^{fluid} denotes the fluid phonon stress intensity factor which can be defined as follows:

$$K_1^{\text{phonon}} = \lim_{x \to a_0^+} \sqrt{\pi(x - a_0)}\sigma_{yy}(\text{x}, 0, t) \tag{7}$$

$$K_1^{\text{fluid}} = \lim_{x \to a_0^+} \sqrt{\pi(x - a_0)}p_{yy}(\text{x}, 0, t) \tag{8}$$

The computational results are dependent upon the ratio σ_0/p_0, while the fluid stress intensity factor is expressed by K_1^{fluid}, which is also a function of time in general, because $K_1^{\text{phonon}}>0$ leads to crack surface opening, and according to our computation, $K_1^{\text{fluid}} < 0$ leads to crack surface closing. The total stress intensity factor is denoted by

$$K_1^{\text{total}} = K_1^{\text{phonon}} + K_1^{\text{fluid}}. \tag{9}$$

Therefore, the value of K_1^{total} is smaller than K_1^{phonon}.

K_1 is the most often used in engineering design parameter in fracture mechanics and, hence, must be understood if we are to design fracture tolerant materials used in bridges, buildings, aircraft, or even bells. Polishing cannot detect a crack. Typically, if a crack can be seen, it is very close to the critical stress state predicted by the stress intensity factor. The magnitude of $\tilde{K}_1$ depends on sample geometry, the size and location of the crack, and the magnitude and the modal distribution of loads on the material.

The normalized elastic dynamic stress intensity factor is defined as

$$\tilde{K}_1(t) = K_1(t)/\sqrt{\pi a_0}\sigma_0 \tag{10}$$

6. Possible criterion of rupture

For soft matter, there is fracture criterion:

$$K_1^{\text{total}} = K_{IC}(T). \tag{11}$$

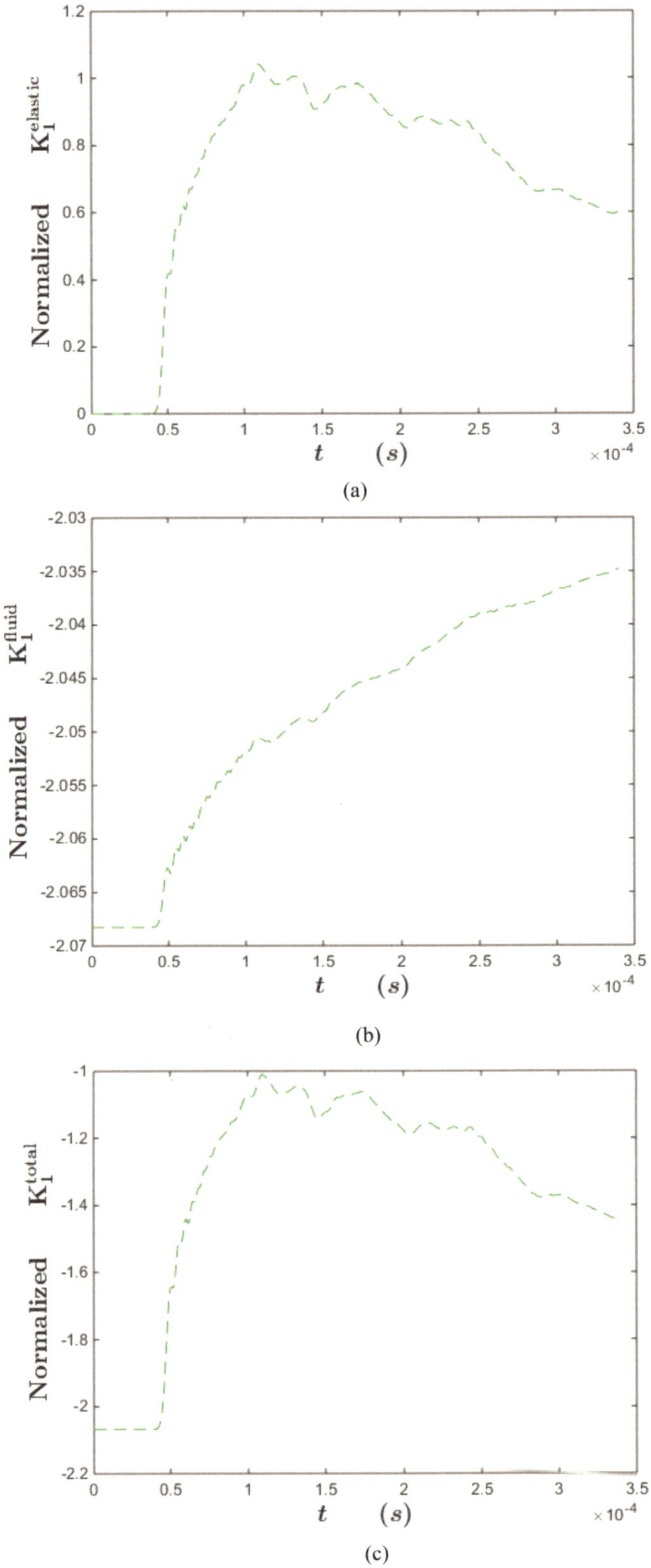

Figure 5.
For soft-matter quasicrystals with 5- and 10-fold symmetries: (a) normalized phonon (elastic) DSIF versus time; (b) normalized fluid DSIF versus time; and (c) normalized total DSIF versus time.

The Griffith theory can be extended to soft matter. If $K_1^{\text{total}}(t) > K_{IC}(T)$ then crack surface to be opening, and crack propagate leads to fracture; otherwise $K_1^{\text{total}}(t) < K_{IC}(T)$ then crack surface to be closing, and crack cannot propagate, the material is in safe. The conclusion is that the physical state around crack tip in soft matter quasicrystals is dependent on the competition of phonon stress and fluid stress.

7. The references for other soft matter

The example given in previous section is only the most preliminary discussion on the probe of rupture of soft-matter quasicrystals. It does not mean any in-depth and exact theory of the problem. The significance of the discussion lies in providing a possible drawing for other categories of soft matter in the quantitative study of rupture problems. The formulation of the present study may be referenced by other branches of soft matter, in which the constitutive laws for themselves are needed, and phason field can be excluded for non-quasicrystalline materials, and so on. With these considerations, the rupture theory for the other categories of soft matter out of the soft-matter quasicrystals may be set up.

8. Conclusion and discussion

From **Figure 5**, we also can see that the real value of fluid DSIF is negative; thus, we can say it causes the crack closed. When comparing (a) with (b), the absolute value of fluid DSIF is larger than the elastic ones; thus, the crack cannot propagate. Then, we can conclude that in the quasicrystal of soft matter, the fluid effect is important, and we should do the study under the hydrodynamics theory.

The equation set of generalized dynamics of soft-matter quasicrystals is complete; it is possible to provide the hydrodynamics study on the mass distribution, deformation, and motion, including rupture analysis of the new phase. In particular, the dynamics stress intensity factor (DSIF) was evaluated, and the computation presents highly stability. In the results, there are two regimes: one is elastic DSIF and the other is fluid DSIF. The former leads to crack surface opening and the latter leads to crack surface closing. We can conclude that whether or not the crack open or close depends on the competition between the elastic DSIF and fluid DSIF. Of course, the theoretical analysis is anxious for the experimental verification currently, which is our attempt work in future.

The above introducing is rough and preliminary, in fact, around the crack tip, due to the stress concentration and high stress grad, nonelastic stress appears, the Griffith theory will not be effective, i.e., the theory based on the stress intensity factor will not be valid, one must carry out nonlinear analysis, e.g. the theory based on the crack tip opening displacement, in which the size of the plastic zone near the crack tip is also considered, refer to Fan [24], which may help us to overcome the difficulty mentioned above. However this is a very difficult question which has not been solved so far.

Acknowledgements

This work is supported by the National Natural Science Foundation of China through grant 11272053.

Conflict of interest

The authors declare no conflict of interest.

A. Appendix

A.1 The final governing equations of generalized dynamics of soft-matter quasicrystals with 10-fold symmetry in three dimension

In the text, Eq. (2) is the planar field form only. In book [24], it provides an explicit form of the equations and lists below soft-matter quasicrystals with 10-fold symmetry:

$$\left.\begin{array}{l}
\dfrac{\partial \rho}{\partial t}+\nabla\cdot(\rho\mathbf{V})=0\\
\dfrac{\partial(\rho V_x)}{\partial t}+\dfrac{\partial(V_x\rho V_x)}{\partial x}+\dfrac{\partial(V_y\rho V_x)}{\partial y}+\dfrac{\partial(V_z\rho V_x)}{\partial z}=-\dfrac{\partial p}{\partial x}+\eta\nabla^2V_x+\dfrac{1}{3}\eta\dfrac{\partial}{\partial x}\nabla\cdot\mathbf{V}\\
+\left(C_{66}\dfrac{\partial^2}{\partial y^2}+C_{44}\dfrac{\partial^2}{\partial z^2}\right)u_x+(C_{12}+C_{66})\dfrac{\partial^2 u_y}{\partial x\partial y}+(C_{13}+C_{44}-C_{11})\dfrac{\partial^2 u_z}{\partial x\partial z}+(C_{11}-B)\dfrac{\partial}{\partial x}\nabla\cdot\mathbf{u}\\
+R\dfrac{\partial}{\partial x}\nabla_1\cdot\mathbf{w}-R\dfrac{\partial}{\partial y}\left(\dfrac{\partial w_x}{\partial y}-\dfrac{\partial w_y}{\partial x}\right)-(A-B)\dfrac{1}{\rho_0}\dfrac{\partial\delta\rho}{\partial x}\\
\dfrac{\partial(\rho V_y)}{\partial t}+\dfrac{\partial(V_x\rho V_y)}{\partial x}+\dfrac{\partial(V_y\rho V_y)}{\partial y}+\dfrac{\partial(V_z\rho V_y)}{\partial z}=-\dfrac{\partial p}{\partial y}+\eta\nabla^2V_y+\dfrac{1}{3}\eta\dfrac{\partial}{\partial y}\nabla\cdot\mathbf{V}\\
+(C_{12}+C_{66})\dfrac{\partial^2 u_x}{\partial x\partial y}+\left(C_{66}\dfrac{\partial^2}{\partial x^2}+C_{11}\dfrac{\partial^2}{\partial y^2}+C_{44}\dfrac{\partial^2}{\partial z^2}\right)u_y+(C_{13}+C_{44})\dfrac{\partial^2 u_z}{\partial y\partial z}+(C_{11}-B)\dfrac{\partial}{\partial y}\nabla\cdot\mathbf{u}\\
-R\dfrac{\partial}{\partial x}\left(\dfrac{\partial w_x}{\partial y}-\dfrac{\partial w_y}{\partial x}\right)-R\dfrac{\partial}{\partial y}\nabla_1\cdot\mathbf{w}-(A-B)\dfrac{1}{\rho_0}\dfrac{\partial\delta\rho}{\partial y}\\
\dfrac{\partial(\rho V_z)}{\partial t}+\dfrac{\partial(V_x\rho V_z)}{\partial x}+\dfrac{\partial(V_y\rho V_z)}{\partial y}+\dfrac{\partial(V_z\rho V_z)}{\partial z}=-\dfrac{\partial p}{\partial z}+\eta\nabla^2V_z+\dfrac{1}{3}\eta\dfrac{\partial}{\partial z}\nabla\cdot\mathbf{V}\\
+\left(C_{44}\dfrac{\partial^2}{\partial x^2}+C_{44}\dfrac{\partial^2}{\partial y^2}+(C_{33}-C_{13}-C_{44})\dfrac{\partial^2}{\partial z^2}\right)u_z+(C_{13}+C_{44}-\mathrm{B})\dfrac{\partial}{\partial z}\nabla\cdot\mathbf{u}-(A-B)\dfrac{1}{\rho_0}\dfrac{\partial\delta\rho}{\partial z}\\
\dfrac{\partial u_x}{\partial t}+V_x\dfrac{\partial u_x}{\partial x}+V_y\dfrac{\partial u_x}{\partial y}+V_z\dfrac{\partial u_x}{\partial z}\\
=V_x+\Gamma_{\mathbf{u}}\left[\left(C_{11}\dfrac{\partial^2}{\partial x^2}+C_{66}\dfrac{\partial^2}{\partial y^2}+C_{44}\dfrac{\partial^2}{\partial z^2}\right)u_x+(C_{12}+C_{66})\dfrac{\partial^2 u_y}{\partial x\partial y}\right.\\
\left.+(C_{13}+C_{44})\dfrac{\partial^2 u_z}{\partial x\partial z}+R\dfrac{\partial}{\partial x}\nabla_1\cdot\mathbf{w}-R\dfrac{\partial}{\partial y}\left(\dfrac{\partial w_x}{\partial y}-\dfrac{\partial w_y}{\partial x}\right)\right]\\
\dfrac{\partial u_y}{\partial t}+V_x\dfrac{\partial u_y}{\partial x}+V_y\dfrac{\partial u_y}{\partial y}+V_z\dfrac{\partial u_y}{\partial z}\\
=V_y+\Gamma_{\mathbf{u}}\left[(C_{12}+C_{66})\dfrac{\partial^2 u_x}{\partial x\partial y}+\left(C_{66}\dfrac{\partial^2}{\partial x^2}+C_{11}\dfrac{\partial^2}{\partial y^2}+C_{44}\dfrac{\partial^2}{\partial z^2}\right)u_y\right.\\
\left.+(C_{13}+C_{44})\dfrac{\partial^2 u_z}{\partial y\partial z}-R\dfrac{\partial}{\partial x}\left(\dfrac{\partial w_x}{\partial y}-\dfrac{\partial w_y}{\partial x}\right)-R\dfrac{\partial}{\partial y}\nabla_1\cdot\mathbf{w}\right]\\
\dfrac{\partial u_z}{\partial t}+V_x\dfrac{\partial u_z}{\partial x}+V_y\dfrac{\partial u_z}{\partial y}+V_z\dfrac{\partial u_z}{\partial z}=V_z+\Gamma_{\mathbf{u}}\left[(C_{13}+C_{44})\left(\dfrac{\partial^2 u_x}{\partial x\partial z}+\dfrac{\partial^2 u_y}{\partial y\partial z}\right)\right.\\
\left.+\left(C_{44}\dfrac{\partial^2}{\partial x^2}+C_{44}\dfrac{\partial^2}{\partial y^2}+C_{33}\dfrac{\partial^2}{\partial z^2}\right)u_z\right]\\
\dfrac{\partial w_x}{\partial t}+V_x\dfrac{\partial w_x}{\partial x}+V_y\dfrac{\partial w_x}{\partial y}+V_z\dfrac{\partial w_x}{\partial z}\\
=\Gamma_{\mathbf{w}}\left[K_1\nabla_1^2w_x+K_4\dfrac{\partial^2 w_x}{\partial z^2}+K_2\dfrac{\partial}{\partial y}\left(\dfrac{\partial}{\partial x}-\dfrac{\partial}{\partial y}\right)w_y+R\dfrac{\partial}{\partial x}\left(\dfrac{\partial u_x}{\partial x}-\dfrac{\partial u_y}{\partial y}\right)-R\dfrac{\partial}{\partial y}\left(\dfrac{\partial u_x}{\partial y}+\dfrac{\partial u_y}{\partial x}\right)\right]\\
\dfrac{\partial w_y}{\partial t}+V_x\dfrac{\partial w_y}{\partial x}+V_y\dfrac{\partial w_y}{\partial y}V_z\dfrac{\partial w_y}{\partial z}\\
=\Gamma_{\mathbf{w}}\left[K_1\nabla_1^2w_y+K_4\dfrac{\partial^2 w_y}{\partial z^2}+R\dfrac{\partial}{\partial x}\left(\dfrac{\partial u_x}{\partial y}+\dfrac{\partial u_y}{\partial x}\right)+R\dfrac{\partial}{\partial y}\left(\dfrac{\partial u_x}{\partial x}-\dfrac{\partial u_y}{\partial y}\right)\right]\\
p=f(\rho)=3\dfrac{k_BT}{l^3\rho_0^3}(\rho_0^2\rho+\rho_0\rho^2+\rho^3)
\end{array}\right\}\quad\text{(A1)}$$

in which $\nabla^2 = \frac{\partial^2}{\partial x^2} + \frac{\partial^2}{\partial y^2} + \frac{\partial^2}{\partial z^2}$, $\nabla_1^2 = \frac{\partial^2}{\partial x^2} + \frac{\partial^2}{\partial y^2}$, $\nabla = \mathbf{i}\frac{\partial}{\partial x} + \mathbf{j}\frac{\partial}{\partial y} + \mathbf{k}\frac{\partial}{\partial z}$, $\mathbf{V} = \mathbf{i}V_x + \mathbf{j}V_y + \mathbf{k}V_z$, $\mathbf{u} = \mathbf{i}u_x + \mathbf{j}u_y + \mathbf{k}u_z$, and $C_{11}, C_{12}, C_{13}, C_{33}, C_{44}, C_{66} = (C_{11} - C_{12})/2$ are the phonon elastic constants, K_1, K_2, K_3, K_4 are the phason elastic constants, R is the phonon-phason coupling constant, η is the fluid dynamic viscosity, Γ_u and Γ_w the phonon and phason dissipation coefficients, and A and B the material constants due to variation of mass density, respectively.

Equation (A1) is the final governing equations of dynamics of soft-matter quasicrystals of 10-fold symmetry in three-dimensional case with field variables $u_x, u_y, u_z, w_x, w_y, V_x, V_y, V_z, \rho$ and p; the amount of the field variables is 10, and the amount of field equations is 10 too; among them are: (A1a) (the first of (A1)) is the mass conservation equation, (A1b)–(A1d) (the second to fourth of (A1)) are the momentum conservation equations or generalized Navier-Stokes equations, (A1e)–(A1g) (the fifth to seventh of (A1)) are the equations of motion of phonons due to the symmetry breaking, (A1h) and (A1i) (the eighth to ninth of (A1)) are the phason dissipation equations, and (A1j) (the tenth of (A1)) is the equation of state, respectively. The equations are consistent to be mathematical solvability, if there is lack of the equation of state, the equation system is not closed, and has no meaning mathematically and physically. This shows that the equation of state is necessary.

The equation set (A1) is the three-dimensional form of the equations, and in our solution, we computed only the two-dimensional form, i.e., the plane field form; in this special case, the 5- and 10-fold symmetry quasicrystals have the same governing equations, but in the three-dimensional case, the equation set (A1) is valid only for the 10-fold symmetry quasicrystals.

Author details

Hui Cheng[1] and Tian-You Fan[2*]

1 School of Mathematics and Physics, Hebei University of Engineering, Handan, P.R. China

2 School of Physics, Beijing Institute of Technology, Beijing, P.R. China

*Address all correspondence to: tyfan2013@163.com

References

[1] Zeng X, Ungar G, Liu Y, Percec V, Dulcey AE, Hobbs JK. Supramolecular dendritic liquid quasicrystals. Nature. 2004;**428**(6979):157-160. DOI: 10.1038/nature02368

[2] Takano A, Kawashima W, Noro A, Isono Y, Tanaka N, Dotera T, et al. A mesoscopic Archimedean tiling having a new complexity in an ABC star polymer. Journal of Polymer Science Part B: Polymer Physics. 2005;**43**(18):2427-2432. DOI: 10.1002/polb.20537

[3] Hayashida K, Dotera T, Takano A, Matsushita Y. Polymeric quasicrystal: Mesoscopic quasicrystalline tiling in ABC star polymers. Physical Review Letters. 2007;**98**(19):195502. DOI: 10.1103/PhysRevLett.98.195502

[4] Talapin DV, Shevchenko EV, Bodnarchuk MI, Ye X, Chen J, Murray CB. Quasicrystalline order in self-assembled binary nanoparticle superlattices. Nature. 2009;**461**(7266):964-967. DOI: 10.1038/nature08439

[5] Fischer S, Exner A, Zielske K, Perlich J, Deloudi S, Steurer W, et al. Colloidal quasicrystals with 12-fold and 18-fold diffraction symmetry. Proceedings of the National Academy of Sciences. 2011;**108**(5):1810-1814. DOI: 10.1073/pnas.1008695108

[6] Cheng SZD. Giant surfactants based on precisely functionalized POSS nano-atoms: Tuning from crystals to Frank-Kasper phases and quasicrystals. In: First Annual Symposium on Frontiers of Soft Matter Science and Engineering, Beijing; 2015. Bulletin of the American Physical Society, 2016

[7] Yue K, Huang MJ, Marson R, He JL, Huang JH, Zhou Z, et al. Geometry induced sequence of nanoscale Frank-Kasper and quasicrystal mesophases in giant surfactants. Proceedings of the National Academy of Sciences. 2016;**113**:1392-1400

[8] Goehring L, Conroy R, Akhter A, et al. Evolution of mud-crack patterns during repeated drying cycles. Soft Matter. 2010;**6**(15):3562-3567. DOI: 10.1039/b922206e

[9] Carbone G, Pierro E, Gorb SN. Origin of the superior adhesive performance of mushroom-shaped microstructured surfaces. Soft Matter. 2011;**7**(12):5545-5552. DOI: 10.1039/c0sm01482f

[10] Xuan Y, Guo X, Cui Y, et al. Crack-free controlled wrinkling of a bilayer film with a gradient interface. Soft Matter. 2012;**8**(37):9603-9609. DOI: 10.1039/c2sm25487e

[11] Haque MA, Kurokawa T, Kamita G, et al. Lamellar bilayers as reversible sacrificial bonds to toughen hydrogel: Hysteresis, self-recovery, fatigue resistance, and crack blunting. Macromolecules. 2011;**44**(22):8916-8924. DOI: 10.1021/ma201653t

[12] Hatton B, Mishchenko L, Davis S, et al. Assembly of large-area, highly ordered, crack-free inverse opal films. Proceedings of the National Academy of Sciences. 2010;**107**(23):10354-10359. DOI: 10.1073/pnas.1000954107

[13] Lazarus V, Pauchard L. From craquelures to spiral crack patterns: Influence of layer thickness on the crack patterns induced by desiccation. Soft Matter. 2011;**7**(6):2552-2559. DOI: 10.1039/c0sm00900h

[14] Tirumkudulu MS, Russel WB. Cracking in drying latex films. Langmuir. 2005;**21**(11):4938-4948. DOI: 10.1021/la048298k

[15] van der Kooij HM, Sprakel J. Watching paint dry; more exciting than

it seems. Soft Matter. 2015;**11**(32): 6353-6359. DOI: 10.1039/C5SM01505G

[16] Martinez CJ, Lewis JA. Shape evolution and stress development during latex-silica film formation. Langmuir. 2002;**18**(12):4689-4698. DOI: 10.1021/la0114833

[17] Tirumkudulu MS, Russel WB. Role of capillary stresses in film formation. Langmuir. 2004;**20**(7):2947-2961. DOI: 10.1021/la0356250

[18] Bohn S, Pauchard L, Couder Y. Hierarchical crack pattern as formed by successive domain divisions. Physical Review E. 2005;**71**(4):046214. DOI: 10.1103/PhysRevE.71.046215

[19] Griffith AA. The phenomena of rupture and flow in solids. Philosophical Transactions of the Royal Society A: Mathematical, Physical and Engineering Sciences. 1921;**221**:582-593. DOI: 10.2307/91192

[20] Fan TY. Mathematical Theory of Elasticity of Quasicrystals and Its Applications. 1st ed./2nd ed. Beijing/ Heidelberg: Science Press/Springer-Verlag; 2010/2016. DOI: 10.1007/978-3-642-14643-5

[21] Fan TY. Poisson bracket method and its applications to quasicrystals, liquid crystals and a class of soft matter. Acta Mechanica Sinica. 2013;**45**(4):548-559. DOI: 10.6052/0459-1879-12-346 (in Chinese)

[22] Fan TY. Equation system of generalized hydrodynamics of soft-matter quasicrystals. Applied Mathematics and Mechanics. 2016; **37**(4):331-344 (in Chinese)

[23] Fan TY. Generalized hydrodynamics for second 2D soft-matter quasicrystals. Applied Mathematics and Mechanics. 2017; **38**(2):189-199 (in Chinese)

[24] Fan TY. Generalized Dynamics of Soft-Matter Quasicrystals—Mathematical Models and Solutions. Beijing/Heidelberg: Beijing Institute of Technology Press/Springer-Verlag; 2017

[25] Landau LD, Lifshitz EM. Theoretical Physics V, Statistical Physics, Part I. 3rd ed. New York: Pergamon Press; 1980

[26] Anderson PW. Basic Notations of Condensed Matter Physics. Menlo Park: Benjamin Cummings; 1984

[27] Lifshitz EM, Pitaevskii LP. Theoretical Physics V, Statistical Physics, Part II. New York: Pergamon Press; 1980

[28] Fan L, Fan TY. Equation of state of structured liquid. In: First Annual Symposium on Frontiers of Soft Matter Science and Engineering; December 12, 2015; Beijing, China; 2015

Chapter 2

Application of J Integral for the Fracture Assessment of Welded Polymeric Components

Zoltan Major, Daniel Kimpfbeck and Matei C. Miron

Abstract

For many demanding applications of engineering plastics, fracture behaviour under various loading conditions is of prime practical importance. It is well known that fracture properties of plastics are significantly affected by the loading rate, temperature and both local and global stress states. The limitations associated with conventional fracture test methods may, at least in principle, be overcome by the use of appropriate fracture mechanical approaches, which properly account for the temperature and rate dependence of the mechanical behaviour of plastics and should provide geometry-independent fracture toughness values. To provide an additional contribution to this application, fracture tests were performed on both 15- and 20-mm-thick bulk-extruded sheets of a polypropylene random copolymer (PP(RC)) and on four different configurations of their welded joints. The fully ductile fracture range was determined by rate-dependent tests on single CT specimens, and fracture toughness values were derived at the peak loads (J_{Fmax} and $CTOD_{Fmax}$). Fracture toughness values were determined for stable crack extension based on the J-Δa and/or $CTOD$-Δa R-curves using single and multiple specimens in terms of various definitions of the crack initiation ($J_{0.2}$, $J_{0.2BL}$ or $\delta_{0.2}$) toughness values. As expected, both methods revealed distinct differences between the bulk materials and the welded joints. These differences were found to depend on the loading rate, the weld configuration and on the data reduction method (J integral or $CTOD$).

Keywords: bulk polymer, welded polymeric structures, elastic–plastic fracture mechanics (EPFM), CTOD, J integral, ductile-brittle transition

1. Introduction

For many demanding applications of engineering plastics, fracture behaviour under various loading conditions is of prime practical importance. In this context it is well known that fracture properties of plastics are significantly affected by the loading rate, temperature and both local (if notches or cracks are present) and global stress states (component geometry or specimen configuration). As a result of the complex effects of these parameters, fracture values determined by conventional test methods (e.g. unnotched and notched Charpy fracture energies) are only of very limited use for material characterisation, especially for engineering design purposes. The limitations associated with conventional testing methods may be overcome by the use of appropriate fracture mechanical approaches, which properly account for temperature and rate dependence of the mechanical behaviour of plastics and should provide geometry-independent fracture toughness values. However, some basic problems should be addressed [1–9]:

- All fracture mechanic methods and their corresponding fracture toughness values have an applicability limit. This limit is associated with the degree of crack-tip yielding and is attended by both theoretical and experimental consequences. These limits are rate- and temperature-dependent for polymers.

- In general, a brittle-to-ductile transition occurs in the fracture behaviour as the test temperature is increased from low to high temperatures. Due to the viscoelastic nature of plastics, this brittle-to-ductile transition depends on the loading rate and also on the local and global stress state.

- True fracture toughness values should be geometry-independent. Several standards and proposals deal with either simple geometry criteria in terms of specimen thickness and ligament length or with a more complicated definition of the local or global specimen constraints [3, 5, 9]. Finally, the geometry dependence may be investigated using various specimen configurations under bending or tensile loading.

- The concepts and data reduction schemes of various fracture mechanics concepts do not support the continuous characterisation of fracture toughness values over a wide loading rate and temperature range.

1.1 Fracture mechanics concepts for various degrees of yielding

In view of the excellent existing literature on various fracture mechanics concepts for different degrees of crack-tip yielding [1–9], no attempt will be made to discuss the subject in great detail. Hence, the following discussion will be restricted to merely referring to the numerous fracture mechanics approaches for monotonic loading.

In a small-scale yielding situation, where linear elastic fracture mechanic (LEFM) concepts may be applied, yielding is limited to a small zone in the immediate vicinity of the crack tip. The relevant crack loading parameters are either the stress intensity factor, K, or the strain energy release rate, G [2, 4, 6]. Crack extension usually occurs in an unstable manner, and the relevant fracture parameters for crack initiation are the critical stress intensity factor, K_{Ic} (also termed fracture toughness), or the critical strain energy release rate, G_{Ic} (the subscript "*Ic*" stands for mode *I* plane-strain conditions). If crack growth occurs in a stable manner, *K*- or *G*-based crack resistance curves (*R*-curves, K vs. Δa or G vs. Δa with Δa being the crack extension) are used.

In a larger-scale yielding situation, where elastic–plastic fracture mechanic (EPFM) concepts may be applied, crack-tip yielding is more extensive but does not spread to the lateral boundary of the specimen prior to the onset of crack growth. The relevant crack-tip loading parameters are the crack-tip opening displacement (*CTOD*) [2, 3] and the *J* integral [2, 3], respectively. Crack growth occurs either by a small amount of stable crack extension followed by unstable rapid crack growth (semi-ductile) or by stable crack extension without any sign of instability (ductile). The relevant fracture parameters for crack initiation are the critical *CTOD* (δ_c) and several definitions of critical *J* integral values (onset of stable fracture, J_Q, J_c or plane-strain fracture toughness, J_{Ic}), respectively. If crack growth occurs in a stable manner, *CTOD* or *J*-based crack resistance curves (*R*-curves) are used. In the latter case, a value for the tearing modulus may also be derived to characterise crack propagation. However, there is still a critical debate about geometry independence and hence applicability of these values.

In a cross-sectional yielding situation, where post-yield fracture mechanics (PYFM) concepts are needed, crack-tip yielding is very extensive and spreads to the lateral specimen boundary ahead of the crack prior to crack extension. The relevant crack-tip loading parameters are derived either from the limit load analysis (LLA) [2, 3] or the essential work of fracture (EWF) concept [4]. In this case specimen failure occurs by plastic collapse of the specimen net section. The characteristic fracture parameters for crack initiation are limit load force values, FLL and specific essential work of fracture values, w_e, respectively. Within the EWF concept, crack extension may additionally be characterised by the specific non-essential work of fracture, W_p. However, geometry independence is also a critical issue here.

1.2 Definition of CTOD

Wells [7, 8] proposed the opening at the crack tip as a measure of the fracture toughness for ductile materials. The critical crack-tip opening displacement (*CTOD*) was related to the stress intensity factor for small-scale yielding. In case of LEFM, the elastic solution for the *CTOD* can still be used:

$$\mathrm{CTOD} = \frac{4}{\pi}\frac{K_I^2}{E\sigma_{ys}} \tag{1}$$

where K_I is the stress intensity factor for mode *I* loading, E is the elastic modulus and σ_{ys} is the uniaxial yield stress of the specific material (for polymers both are rate- and temperature-dependent).

The direct measurement of *CTOD* was considered to be difficult until the optical full-field displacement and strain analysis methods were frequently applied. According to the related standard (ASTM E1290), the *CTOD* is estimated from crack opening displacement (*COD*) measurements at the specimen surface using a clip-gauge extensometer. The *CTOD* (δ) can also be separated into two components and calculated from $\delta = \delta_{el} + \delta_{pl}$. While the elastic component is equal to Eq. (1), the plastic component, δ_{pl}, is obtained from an analysis that assumes a rotation point near the centre of the ligament and can be calculated [9]. The rapid development and frequent use of full-field optical methods for fracture analysis opened new options for the direct *CTOD* measurements as it is described in [10–12]. The resolution of the entire optical test set-up must be sufficient for determining proper *CTOD* values [10–12].

1.3 *J* integral definition

The theoretical *J* integral concept was proposed by Cherepanov (1967) and Rice (1968) as a path-independent integral and was used originally as a measure of the intensity of elastic–plastic crack-tip fields [13–17]:

$$J = \int_C \left(W dy - T \cdot \frac{\partial u}{\partial x} ds \right) \tag{2}$$

where W is the strain energy density, T is the traction vector, u is displacement vector and $\partial u/\partial x$ is the deformation gradient. The *J* integral defined by Rice in terms of a line-independent path integral cannot be easily measured experimentally. For simple bend-type fracture specimens, however, a straightforward analysis has been developed to relate J to the area under the load versus load point displacement record [9].

For ductile fracture, the toughness can be measured either as a single-point value or as a multi-point function in a resistance curve format and is often characterised by the J integral [18, 19]. Single-point values are based on selected parts of the load–load-point displacement curves. The energy values are determined up to these selected values, and the J integral values are calculated based on the simplified definition of the J integral. In this case, various specimen geometries (SENB and CT) can be tested as single specimen are fractured at a number of different loading rates yielding corresponding fracture parameters for these loading rates. In addition to the plain-sided specimens, in order to reduce the plasticity at the ligament, side-grooves may also be applied (SENB-SG or CT-SG). A data reduction scheme based on numerous values of the J integral was applied, and corresponding values for J were determined as follows:

$$J = \frac{\eta(a/W)U}{B(W-a)} \tag{3}$$

where U is the fracture energy up to the relevant force (F_m for J_m or J_c and F_u for J_u), $\eta(a/W)$ is a crack length-dependent geometry factor, B is the specimen thickness (corresponding to B_{SG}, the net thickness without side-grooves for side-grooved specimen), W is the specimen width and a is the crack length. While J_c is applicable to characterise quasi-brittle failure behaviour (quasi-linear load–displacement behaviour with a sharp load drop at the point of fracture), J_m and J_u refer to J integral definitions in which the peak loads, F_m, and the ultimate loads, F_u, respectively, are used in the data reduction procedure for materials with non-linear displacement traces and a certain amount of stable crack extension prior to ultimate specimen failure. Details for the determination of the geometry factor, $\eta(a/W)$, are given in [2, 3, 5, 6]. All fracture energy values were corrected by the indentation energy according to a procedure described in [20].

Usually, a J integral-based resistance curve (i.e. a J-R-curve) is used to describe a ductile material resistance against stable crack initiation, stable crack growth and tearing instability. In this second case, load–displacement curves can be recorded up to different displacements using several specimens to generate crack extension curves (R-curves). These specimens will be broken in liquid nitrogen, and the crack length related to stable crack growth is determined on the fracture surface.

These tests can be performed according to proposals by ASTM [9, 18] and ESIS [21] in terms of the values of $J_{0.2}$, $J_{0.2BL}$. In the first case, $J_{0.2}$ values are determined, which characterise the fracture resistance at 0.2 mm of the crack extension. Here, the blunting of the ductile crack-tip deformation is not considered. The consideration of the blunting is rather typical for metals. As the blunting line depends on the yield stress of the material, it will inherently be dependent on the loading rate and temperature for polymers.

For higher testing rates (over 1–10 mm/s depending on the material), a special experimental procedure was developed to determine the R-curves. In these tests the actuator movement was stopped at various predefined deformations prior to total specimen fracture, and the load and displacement data were recorded. The crack extension-dependent J integral values were determined according to data reduction procedures described elsewhere [22, 23].

Because of their effectiveness in measuring toughness, the J integral and J-R-curve have become the most important material parameters in elastic–plastic fracture mechanics and have been applied widely in practical engineering. Simultaneously, the *CTOD* values can also be used in *R*-curves. Such fracture toughness values based on the J-Δa and/or *CTOD*-Δa *R*-curves in terms of various definition of the crack initiation ($J_{0.2}$ $J_{0.2BL}$ or δ_c) may serve as a basis for material characterisation, performance evaluation and quality assurance.

A number of different research groups dealt with various problems of the experimental determination and the application to polymers. It must be highlighted the detailed and comprehensive polymer science-based work of Grellmann and Seidler [22–24] on the field of the determination of structure–property relationships using *J* integral and *CTOD* values for a number of polymeric materials. In addition, this group contributed significantly to the development of the experimental determination of *J* and *CTOD* values at high loading rates. Furthermore, the working group of ESIS TC4 [25–28] provided lot of contributions to the experimental determination of *J* values for polymers.

Elastic–plastic fracture mechanics parameters can also be used for structural damage tolerance assessment, fitness-for-service evaluation, residual strength analysis and structural integrity management for various engineering components and structures, such as pressure vessels and piping in nuclear power plants, onshore and offshore petrochemical vessels and tanks, pipelines in oil and gas industries and aircraft structures [19].

Some fire-service vehicles are equipped with tanks for quenching liquids made from extruded polypropylene copolymer (PP-RC) sheets. Several hundred metres of extrusion welding joints are used in these structures. Due to the geometrical constraints, various weld configurations are used in these tanks. Although, the quality of these welded joints was significantly improved, a large number of different defects and imperfections were recognised in previous experimental investigations. The bulk polymeric material revealed pronounced ductility even at high strain rates and low temperatures. Hence, in addition to the conventional stiffness and strength analysis of the tank, this welded structure was considered as an ideal model component for elastic–plastic fracture mechanics (EPFM) investigations.

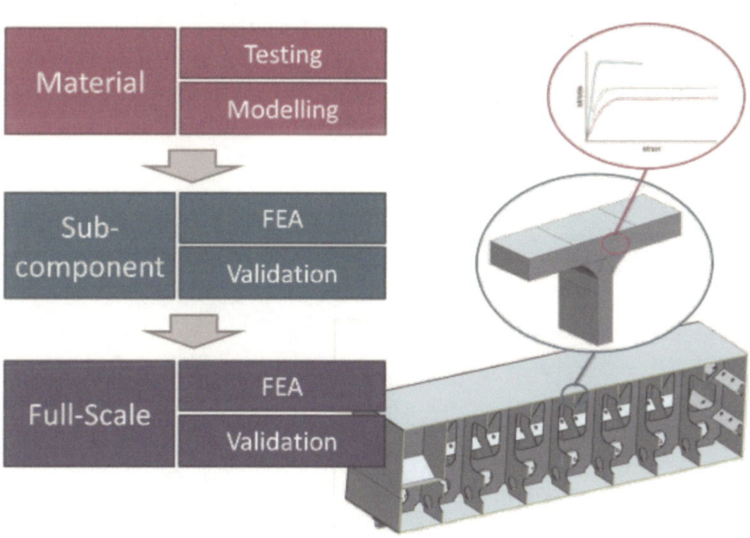

Figure 1.
Schematic representation of the multi-scale approach for the specific welded polymeric structure. Laboratory specimen level, stress–strain curves (top), welded T-joint (mid) and the entire welded polymeric structure (bottom).

The conventional investigations are summarised in the thesis of Distlbacher [29]. This thesis was structured according to the scale level of analysis, and it is illustrated in **Figure 1**. The complex structure of the tank was broken down into smaller testable subcomponents. The subcomponents represent the various types of join design configurations used in the tank.

2. Basic considerations

2.1 Experimental techniques of EPFM

2.1.1 Multiple-specimen method of J-Δa curve

The different specimens are loaded up to different displacement levels. These displacements generate different lengths of stable crack growth. The *J* integral values are calculated using the area of the load–displacement curves up to the corresponding displacement levels. The specimens are then immersed into liquid nitrogen and subsequently broken under impact loading. The stable crack growth is measured on the fracture surface as an average of the crack front from the side to the mid of the specimens. The conventionally calculated *J* integral values (energy and geometry factor) can be corrected for large cracks using the *Δa* values measured. The schematic explanation of the multi-specimen method for determining *J-R*-curves is shown in **Figures 2** and **3**:

$$J = \frac{\eta_{el} U_{el}}{B(W-a)} + \frac{\eta_{pl} U_{pl}}{B(W-a)} \left\{ 1 - \frac{(0.75\eta_{el} - 1)\Delta a}{W-a} \right\} \tag{4}$$

The determination of crack resistance curves using multiple specimens is time-consuming and laborious, at least 8–10 specimens should be loaded up to different load levels (displacements). For a statistical evaluation, even more specimens are needed. Hence, there are several efforts to determine reliable *J-Δa* curves using only a single specimen. For polymeric materials, the detection of the stable crack growth is not always an easy task and the specimen preparation as well as the microscopy could be extremely time-consuming. However, this method allows for the reliable determination of critical *J* values with and without blunting line.

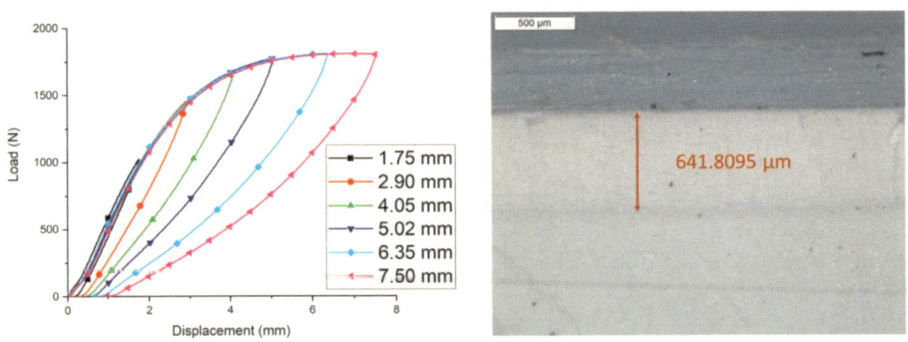

(a) (b)

Figure 2.
Schematic explanation of the multi-specimen method for determining J-R-curves, (a) load–displacement curves at different displacement levels for the determination of first the energy (U_1 (1.75 mm) to U_{max} (7.5 mm)) and then the J integral values, (b) fracture surface with razor blade notch and stable crack growth (a_i = 0.642 mm).

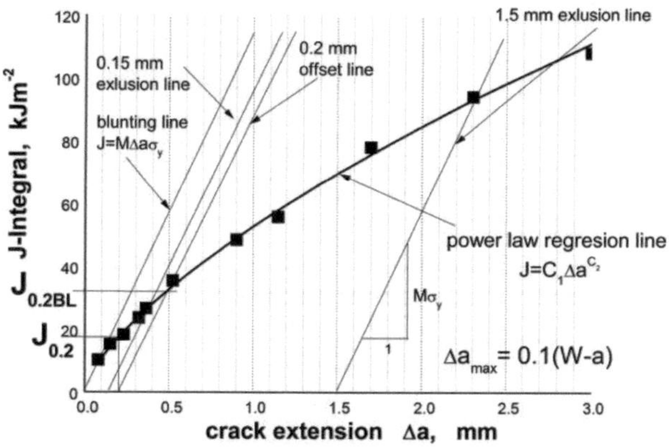

Figure 3.
Construction of the R-curve, J-Δa values measured and exponential fit curve (R-curve); definition of blunting line and the corresponding $J_{0.2BL}$ value and definition of $J_{0.2}$ value. (remark: For polymeric materials the blunting line is loading rate- and temperature-dependent, that is, a unique blunting line belongs to every specific R-curve).

2.1.2 Single-specimen methods of J and J-Δa curve

- Single parameter
  - Derivation of J values from the force-displacement curve measured and set J_c values at a critical force (F_{max}).
- Single specimen with load separation principle
  - Sharobeam and Landes presented an experimental procedure based on the load separation principle to construct the material resistance curve of an elastic–plastic material [25, 30].
- Single specimen with semi-cyclic loading
  - Semi-cyclic loading with increasing displacement used. Loading-unloading curves are measured, and compliance method is applied for determining Δa values [9, 19].
- Single specimen with FFSA
  - Novel experimental techniques offer also new options for determining J-Δa curves using a single specimen. Optical full-field deformation measurement (FFSA) with digital image correlation for determining crack-tip opening displacement ($CTOD$) and crack growth (Δa) simultaneously was used by several authors recently [10–12] and also in this study.

2.2 Application of EPFM methods for polymers

As it was mentioned before, the application of J integral methods for polymeric materials is focused mainly on material characterisation, quality assurance and on the determination of structure–property relationships. There is hardly any application of fracture mechanics tools for structural integrity management for various engineering components of engineering polymers. The majority of unfilled components are produced as thin-walled injection-moulded components for mass applications. The application of complex fracture mechanics methods for these components is too expensive, and the risk of such components to fail does not necessitate their application. On the other hand, continuous fibre-reinforced composites reveal predominantly a linear elastic behaviour. Linear elastic fracture mechanics methods both in terms of stress intensity factor *K*, but even more frequently in terms of strain energy release rate *G*, are widely used for these components. In spite of a high number of investigations, there are still open questions with regard to LEFM; the additional complexity of elastic–plastic method is not rewarded by the users. *J* integral values are, however, often calculated in numerical simulations [31, 32] but also predominantly in the linear elastic deformation regime.

It was recognised as an appropriate occasion to apply EPFM methods for large thickness welded polymer structures. These structures were investigated in a contract research project with conventional strength analysis methods for a company partner. Although the majority of these results are confidential, a larger number and high-quality material data were generated:

1. In terms of loading rate- and temperature-dependent true and nominal stress–strain curves on laboratory specimen level

2. Strength analysis of various weld configurations on subcomponent level

3. Stiffness analysis of the entire structure along with the calculation of local stress concentrations

Hence, in addition to the conventional stiffness and strength analysis of the tank, this welded structure was considered as an ideal model component for elastic–plastic fracture mechanics investigations along with some methodological issues.

Based on the above considerations, the main objectives of this paper are:

- To investigate the application of elastic–plastic fracture mechanics methods in terms of *J* integral and *CTOD* to a bulk (unfilled) polypropylene random copolymer (PP-RC) and its welding joints

- To determine crack resistance curves (*R*-curves) along with loading rate-dependent critical fracture toughness values for the bulk polymer and for four configurations of welded joints

3. Experimental

The specific material that was used for this fracture mechanics analysis in this study is a Polystone P-RFT® (Röchling Engineering Plastics SE & Co.KG, Haren, Germany) further termed as Polystone PP. Polystone PP is a compound material, with a polypropylene random copolymer (PP(RC)) as a base material. Due to its high strength, high weldability and excellent chemical and corrosive resistance, it is mainly used in chemical engineering and tank building and reveals the following basic properties at room temperature: Young's modulus E = 1534 MPa; Poisson's ratio $\nu = 0.42$; yield stress $\sigma_y = 30$ MPa; and corresponding yield strain $\varepsilon_y = 4.0\%$ [29].

Thick extruded neat PP(RC) sheets with a nominal thickness of 15 and 20 mm were provided by the company for these investigations. Compact tension (CT) specimen was manufactured using these plates. The CT specimens were tested under monotonic loading conditions over a loading rate range of 1 to 1000 mm/s.

The geometry of the CT specimens, (a) dimensions, (b) with weld and (c) bulk material, is shown in **Figure 4**.

These extruded Polystone PP sheets were welded by extrusion welding using the same material for welding filament at the company partner under controlled conditions. The various weld configurations used here are shown in **Figure 5**. Four different weld seam geometries, square butt welded joint, St; single V butt weld joint, VN-2sp; double V butt weld joint single-sided, 2VN-1sp; and double V butt weld joint double-sided, 2VN-2sp, and two different plate thicknesses 15 and 20 mm were used.

These weld configurations are used in a real welded structure (polymer tank). It must be emphasised here, however, that all four welding configurations have only model character in this study and represent an average of all possible welding quality regarding operator and welding parameters. These welded plates were selected randomly from a larger set of welded plates which were produced for quality assurance purposes. Selected defects in welds are shown in **Figure 6**. These defects range from imperfections (not welded regions) to defects (voids in the weld seam). Polymer welding is frequently used in many practical applications and can be considered as secondary technology for polymer processing. It must be emphasised here, however, that a polymer weld reveals different properties than a usual metal weld. Neither the importance of the welding nor the relative quality compared with the bulk material is at the same level. Our structure represents, even more, an exception for structural design of polymeric structures.

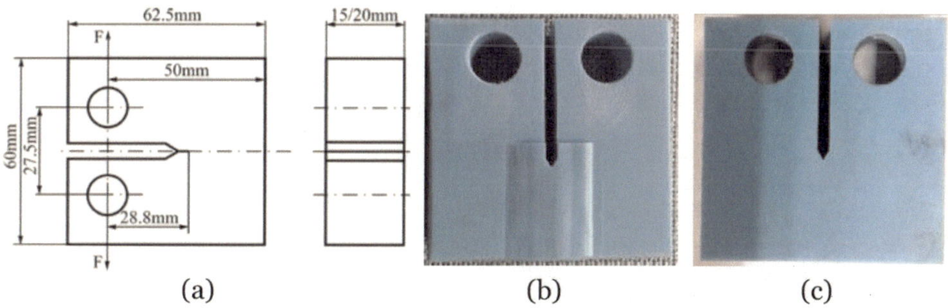

Figure 4.
Geometry of the CT specimens, (a) dimensions, (b) with weld and (c) bulk material.

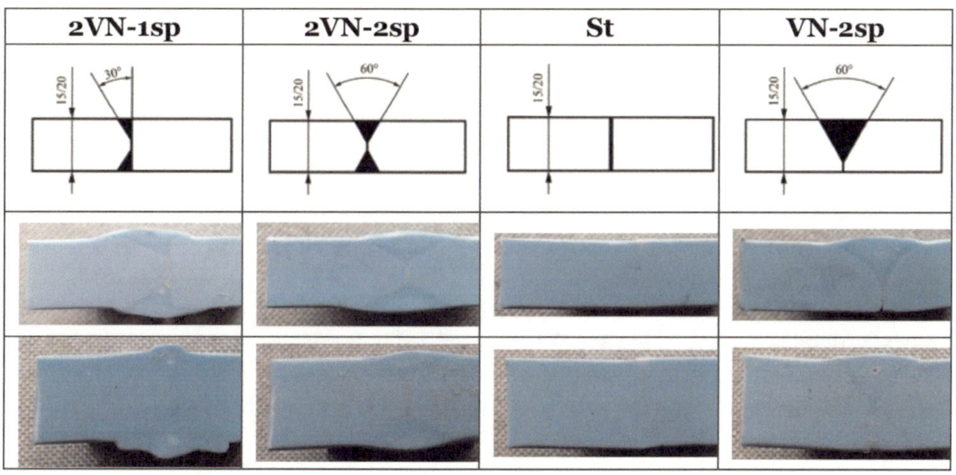

Figure 5.
The different weld configurations used in this study.

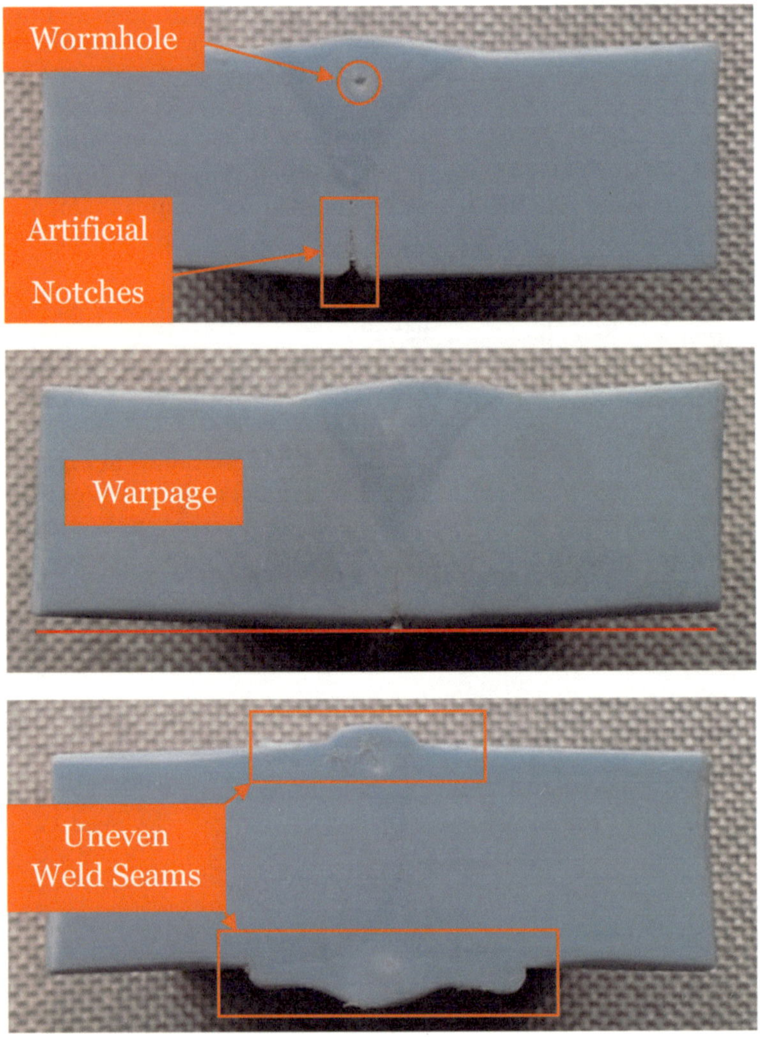

Figure 6.
Detection of various defects in welded polymeric joints.

Schematic representation of the single-specimen method using semi-cyclic loading with increasing displacement along with the fracture surface is shown in **Figure 7**. J integral values are calculated based on the area below the load–displacement curves for every single loading cycle.

The schematic representation of the optical deformation and strain measurement at the vicinity of the crack tip of the CT specimen are shown in **Figure 8**. A region of interest (ROI) was selected from the entire CT specimen (see **Figure 4**). Local crack-tip deformation ($CTOD$, δ_t) and stable crack growth (Δa) can be detected simultaneously.

The data reduction scheme of the single specimen experiments with optical measurements is shown in **Figure 9** (top, force-displacement curves; middle, $CTOD$-displacement curves; and bottom, stable crack growth Δa-displacement curves). The determination of fracture toughness in terms of the following parameters was conducted:

- Critical single J and $CTOD$ values at the maximum force on the force-displacement curve: J_{Fmax} and $CTOD_{Fmax}$

- Critical stable crack initiation values according to the ASTM 1737 proposal using the corresponding blunting lines on the J-Δa curves: $J_{0.2BL}$

- Determination of $J_{0.2}$ values at approximately 0.2 mm crack initiation on the combined force-displacement and Δa-displacement curves

The further obvious advantage of this method for practical fracture assessment is the concurrent determination of *CTOD* and Δa values. There is a long debate about the selection and applicability of both EPFM methods. While *CTOD* is solely a geometrical analysis of the deformation at the vicinity of the crack tip, J integral allows for a more detailed analysis of the stable crack initiation and crack growth process. It involves the elastic–plastic material model at the vicinity of the crack tip. Ductile fracture is associated with large-scale crack-tip yielding and with extensive

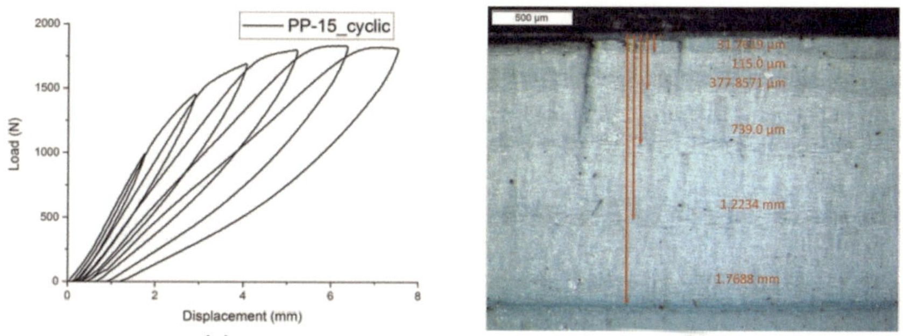

(a) (b)

Figure 7.
Single-specimen method using CT specimen: (a) load–displacement curves of displacement controlled semi-cyclic loading of the specimen for increasing deformations and (b) fracture surface of the CT specimen along with the crack growth marks and the relevant stable crack extension values.

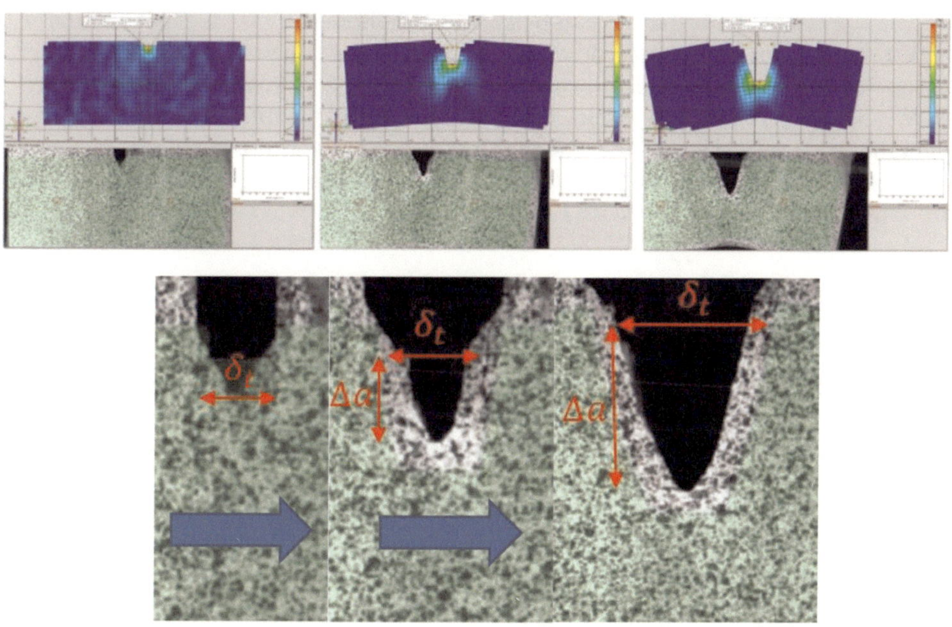

Figure 8.
Schematic representation of the optical deformation and strain measurement at the vicinity of the crack tip of CT specimens (see **Figure 4***) (top FFSA images). Detection of the crack-tip deformation (CTOD, δ_t) and stable crack growth (Δa) in the ROI (bottom images).*

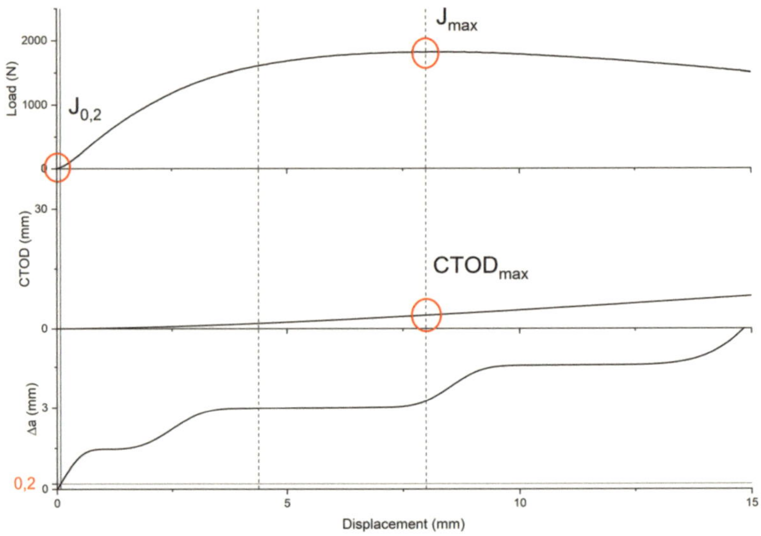

Figure 9.
Data reduction of the single-specimen experiments with optical measurements: Force-displacement curves (top), CTOD-displacement curves (mid) and stable crack growth Δa-displacement curves (bottom).

crack-tip blunting. The blunting of the crack tip can be characterised by the relationship $CTOD$ vs. Δa:

$CTOD > 2\Delta a$, super blunting.

$CTOD = 2\Delta a$, circular blunting and.

$CTOD < 2\Delta a$, less blunting, tendency for unstable crack growth.

Furthermore, the kinetic of the crack growth may be characterised by the actual balance between the values of $CTOD$ and Δa, and based on the actual ratio of these values, the continuity/discontinuity of the crack extension can also be assigned.

The 0.2 mm crack initiation can be considered, however, as an experimentally uncertain limit for the macroscopic image-based single-specimen method. Neither the optical resolution of typical camera systems nor the visibility of the near crack-tip area makes a reliable determination of these values possible. But as neither enough resources were available for the time-consuming multi-specimen method nor a microscope system implemented [22–24] on the test machine was available, the single-specimen method with FFSA was used for the welded joints.

4. Results

Single-specimen tests using the load–displacement curves for determining J values were performed with 20 mm bulk C(T) specimens at loading rates of 1, 10, 100 and 1000 mm/s. The ductile/(semi)-brittle transition was characterised by the appearance of the fracture surfaces along with single J_{Fmax} values.

Load displacement curves of 20-mm-thick bulk CT specimens at various loading rates are shown in **Figure 10**. It was speculated previously that the 20-mm-thick specimens reveal a more pronounced ductile-brittle transition and the transition will occur at lower loading rates than for 15-mm-thick specimens. A clear sign of the ductile-semi-brittle transition is visible on both the curves and fracture surfaces. Hence, the ductile regime was selected for further detailed investigations.

Comparison of the J-Δa data points for multiple-specimen method vs. single-specimen methods (monotonic and cyclic) is seen in **Figure 11**. The three R-curves

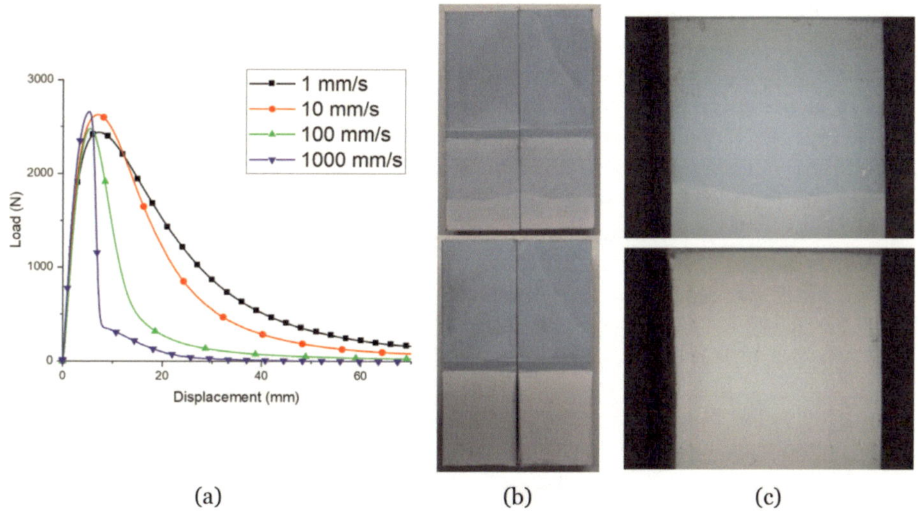

Figure 10.
(a) Load–displacement curves of 20-mm-thick bulk CT specimens at various loading rates, (b) macro images of tested 20-mm-thick bulk CT specimens at 1 mm/s (bottom) and 1000 mm/s (top) (c) along with fracture surfaces at 1 mm/s (bottom) and 1000 mm/s (top).

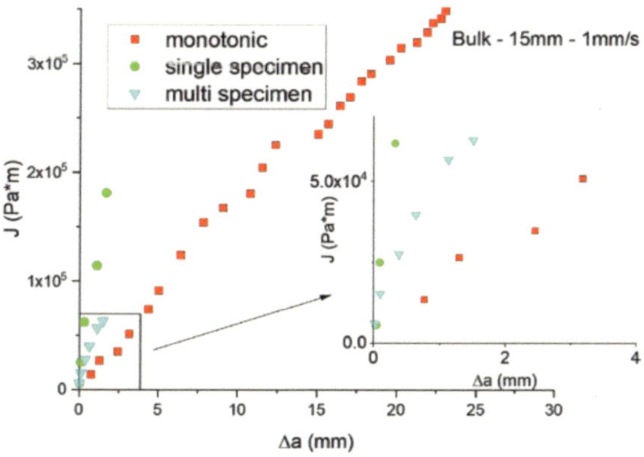

Figure 11.
Comparison of the J-Δa data points for multiple-specimen method vs. single-specimen methods (monotonic and cyclic)—Single-specimen monotonic loading with FFSA (red squares), single-specimen semi-cyclic loading (blue triangles) and multi-specimen (green circles) of a 15-mm-thick PP-bulk specimen obtained at 1 mm/s.

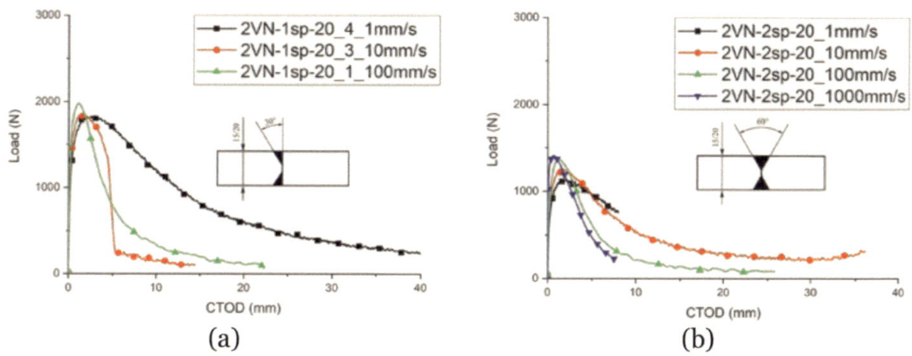

Figure 12.
Load–CTOD curves of 20-mm-thick welded CT specimens at various loading rates, (a) 2VN-1sp and (b) 2VN-2sp weld configurations.

revealed a sufficient concordance but also clear differences. These differences can be assessed as the experimental evidence of the methods under controlled experimental conditions. We apprehended previously that the optical crack length measurement at the specimen side will not reveal sufficient quality at low crack extension levels. The single-specimen method with FFSA can tendentially be used only for longer crack extension. Nevertheless, for a fast screening, it is a convenient and stable method and exhibits a most conservative measure of the crack resistance (lowest J values for given crack length). As the amount of specimen and the

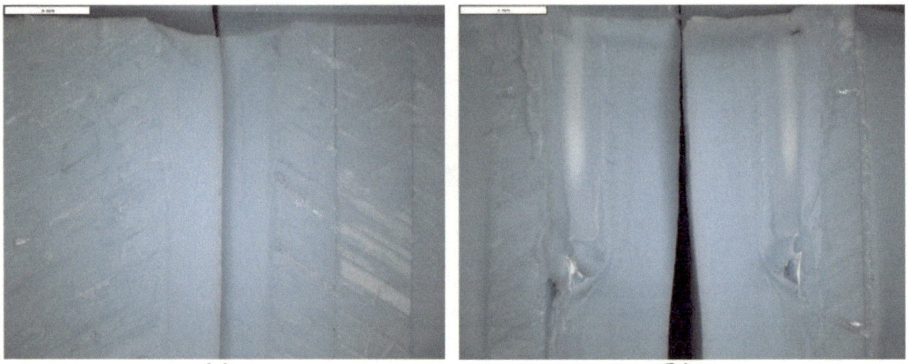

(a) (b)

Figure 13.
Corresponding fracture surfaces at 1 mm/s (a) 2VN-1sp and (b) 2VN-2sp weld configurations. A distinct welding defect is also shown in figure (b).

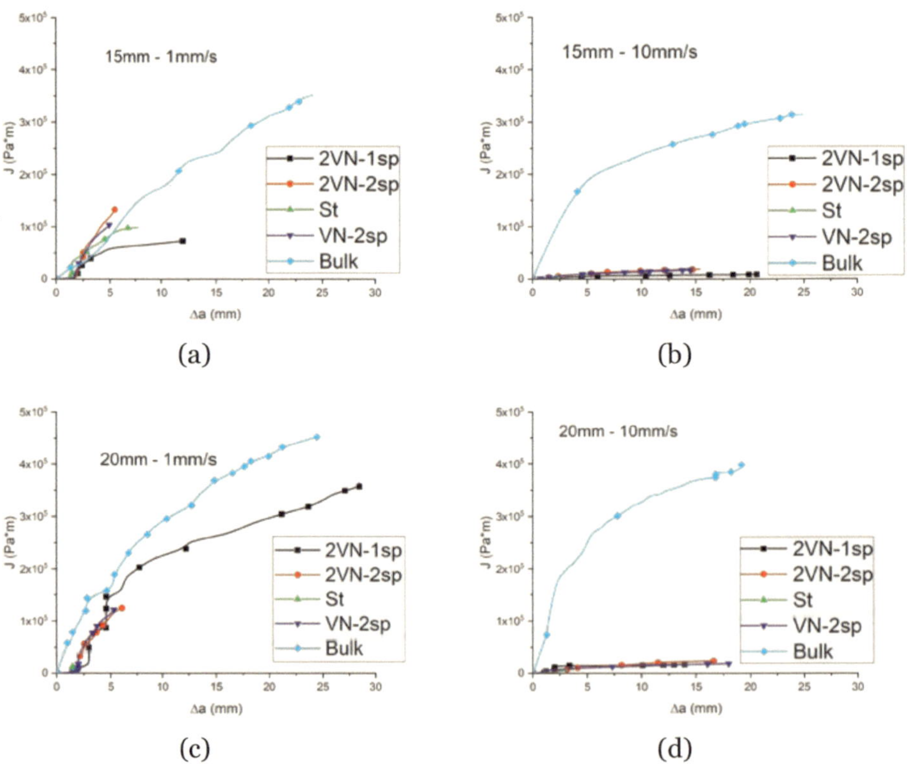

Figure 14.
J-Δa (R) curves for bulk and for welded material at a loading rate of 1 and 10 mm/s for 15- and 20-mm-thick CT specimens. (a) 1 mm/s, 15 mm thickness, (b) 10 mm/s, 15 mm thickness, (c) 1 mm/s, 20 mm thickness and (d) 10 mm/s, 20 mm thickness.

available laboratory resources were also limited, the welded specimens in the second part of the study have only been analysed by the single-specimen FFSA method.

Single-specimen tests using the load–displacement curves for determining the *J* values were also performed with 15 and 20 mm welded CT specimens at loading rates of 1, 10, 100 and 1000 mm/s. Load–*CTOD* curves of the 20-mm-thick welded CT specimens at various loading rates; (a) 2VN-1sp and (b) 2VN-2sp weld configurations are shown in **Figure 12**. The corresponding fracture surfaces are shown in **Figure 13**. A distinct defect is also seen in the mid of the 2VN-2sp weld.

The load-*CTOD* curves of the welded joints revealed ductile fracture at the testing rate range investigated. While a weak sign of ductile-brittle (semi-brittle) transition was observed above 100 mm/s for 2VN-1sp, all 2VN-sp welded polymeric joints revealed stable tearing.

The fracture surfaces of the broken specimens revealed similar behaviour as the load-*CTOD* curves. In spite of the defects, the two exemplary fracture surfaces reflect ductile fracture at low loading rate. This behaviour was assessed as a kind of damage tolerance at least in the specific loading rate and temperature range.

J-Δa (*R*) curves for bulk and for welded material at a loading rate of 1 mm/s and 10 mm/s for 15- and 20-mm-thick CT specimens are shown in **Figure 14a–d**.

CTOD-Δa (*R*) curves for bulk and for welded material at a loading rate of 1 and 10 mm/s for 15- and 20-mm-thick CT specimens are shown in **Figure 15a–d**.

The difference of the *J-R*-curves between the bulk and the welded joints was found significantly rate-dependent. While at 1 mm/s, rather a small difference was

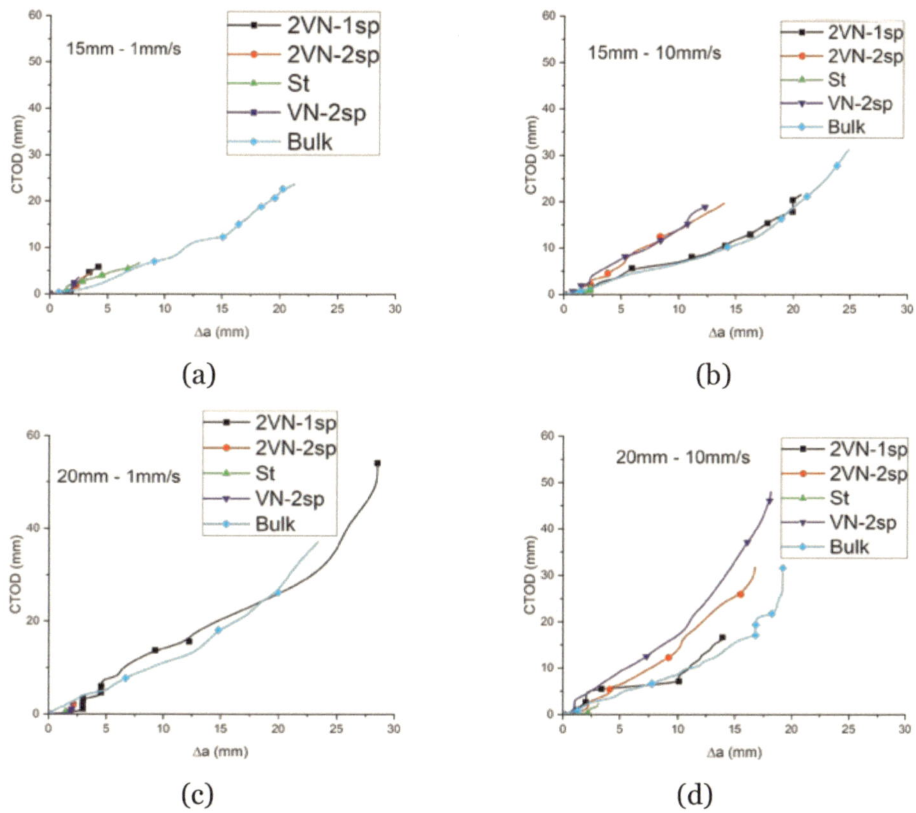

Figure 15.
CTOD-Δa (R) curves for bulk and for welded material at a loading rate of 1 and 10 mm/s for 15- and 20-mm-thick CT specimens. (a) 1 mm/s, 15 mm thickness, (b) 10 mm/s, 15 mm thickness, (c) 1 mm/s, 20 mm thickness and (d) 10 mm/s, 20 mm thickness.

observed for both thicknesses; a significantly larger difference was obtained for 10 mm/s. The welding joints revealed at 10 mm/s a distinct tendency for an embrittlement. The temporary and local softening of PP(RC) and defects seem to prevent the material from acting ductile and instead enforce brittle behaviour under loading. This tendency can already be observed at loading rates as low as 10 mm/s, whereas the bulk material shows no such behaviour at these loading rates. While the bulk polymer by trend has shown *CTOD* values in the range of $2\Delta a$, the welded joints revealed values $CTOD < 2\Delta a$. Hence, while the bulk has revealed distinct blunting, the welding joints showed less blunting prior to crack growth.

Smaller differences between the bulk and weld were recognised based on the *CTOD-R*-curves. In some cases, the welded material revealed similar *CTOD* values than the bulk. A possible explanation for these results is that the razor blade pre-crack was positioned in the bulk, either partly or fully (**Figure 16**).

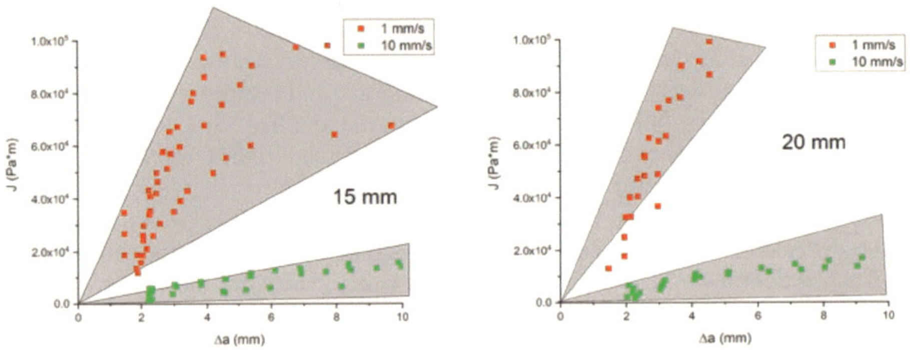

(a) (b)

Figure 16.
Comparison of the general behaviour tendency for different loading rates for the specimen of all different weldments in terms of J-Δa values for the specimen of (a) 15 mm and (b) 20 mm thickness.

	J_{Fmax}, kJm^{-2} 15 mm–1 mm/s	J_{Fmax}, kJm^{-2} 15 mm–10 mm/s	J_{Fmax}, kJm^{-2} 20 mm–1 mm/s	J_{Fmax}, kJm^{-2} 20 mm–10 mm/s
2VN-1sp	15	2	75	6.8
2VN-2sp	41	4.1	40	4.2
St	44	2.8	6.6	13
VN-2sp	30	3.2	3.3	40
Bulk	64	62	86	91.9

Table 1.
J_{Fmax} values of the bulk PP(RC) and welded joints.

	CTOD, mm 15 mm–1 mm/s	*CTOD*, mm 15 mm–10 mm/s	*CTOD*, mm 20 mm–1 mm/s	*CTOD*, mm 20 mm–10 mm/s
2VN-1sp	1.4	1.2	3.4	1.8
2VN-2sp	2.3	2.1	2.6	1.9
St	2.3	1.5	2.3	0.5
VN-2sp	2.3	1.9	1.8	2.4
Bulk	2.2	0.6	2.4	1.1

Table 2.
$CTOD_{Fmax}$ values of the bulk PP(RC) and welded joints.

The 2VN-1sp weld configurations revealed always high *CTOD* values. Based on the above curves, single fracture toughness values, J_c in terms of single J_{Fmax} values, J-Δa curves, as well as $CTOD_{Fmax}$ values and $CTOD$-Δa curves were derived. While the determination of the J_{Fmax} and $CTOD_{Fmax}$ values was easy and reveal sufficient quality for the material characterisation, the determination of $J_{0.2}$ values was not possible for the welded joints using the single-specimen FFSA technique. The bulk $J_{0.2}$ values (2.6–8.2 kJm^{-2}) are in a similar range as literature values [20, 22, 30]. The following two **Tables 1** and **2** and contain these values for the four different weld configurations.

5. Conclusion

Fracture tests were performed on both 15- and 20-mm-thick bulk-extruded sheets of a polypropylene random copolymer and on their welded joints. After the set of the fully ductile failure regime with regard to the loading rate at room temperature, fracture toughness values were determined both in terms of various single *J* integral and *CTOD* values and in terms of crack resistance (J-Δa and $CTOD$-Δa *R*-curves). As expected, the polymer investigated revealed a distinct rate dependence over the entire loading rate range investigated. The difference in the fracture behaviour between the bulk and the welded joints was found to be significant in terms of *J* integral-based fracture parameters (J_{Fmax} and J-Δa curves), and it was also rate-dependent. While at 1 mm/s, rather a small difference was observed; significantly larger difference was obtained at 10 mm/s. Rate-dependent single *F-s* curves for welded joints also support (see **Figure 12a**) this observation. The change of the character of the *F-s* curves in the post-maximum range indicates a smooth ductile to semi-ductile transition. The welded joint reveals constrained deformation around the crack tip. $CTOD$-Δa curves of the welded joints revealed only slightly higher values than the bulk values at 1 mm/s, and also a smaller stable crack extension was observed. The difference was larger at 10 mm/s; however, the 2VN-1sp weld revealed similar curves than the bulk. The welded polymers can exhibit around the welded joints somewhat lower modulus values (10–15%) [29], that is, at constant displacement larger *CTOD* was expected.

The single-specimen method with FFSA is a suitable fast screening method for characterising the overall fracture behaviour of polymeric test specimens but can practically be used only for longer crack extension. The construction of *J-R* and *CTOD-R*-curves was possible for both the bulk and for the welding joints using the single-specimen FFSA method. While the resolution was sufficient for bulk materials also for determining $J_{0.2}$ values, only J_{Fmax} and $CTOD_{Fmax}$ values were determined in sufficient quality for the welded joints of the PP(RC) polymer. The determination of the onset of the stable crack extension and the assignment of $J_{0.2}$ or $J_{0.2BL}$ values is uncertain. The obvious advantage of this method for practical fracture assessment is the concurrent determination of *CTOD* and Δa values. The continuity of the stable crack extension can be properly characterised, and the method provides information about the ability for stable tearing of the cracked specimen.

The simplified energy-based calculation scheme for determining fracture toughness values in terms of critical *J* integral values can be applied without any further consideration for material characterisation and for structure–property relationships. The essential use of fracture toughness parameters is in the fracture assessment and dimensioning of components. The usual fracture mechanics-based dimensioning of a component can be described by the simple relationship below:

$$J(\mathrm{a}, \mathrm{W}, \sigma) \Leftrightarrow J_c(\mathrm{T}, \mathrm{d}\varepsilon/\mathrm{dt}, \mathrm{B}) \quad (5)$$

where J is the non-linear measure of the crack-tip stress intensity (loading) and J_c is the fracture toughness in EPFM. While J depends on the crack length, the relevant size of the components and on the remote stress (on the material model, thus indirectly on temperature and loading rate), J_c depends generally on the loading rate and temperature as well as on the specimen thickness (geometry). While the majority of $J_{0.2}$ values determined for the bulk were classified as J_{Ic} values, the other values can only be interpreted as apparent fracture toughness, J_c. Due to the more strict criteria for fracture instability, the welded joints revealed rather apparent fracture toughness values. Moreover, due to the high data scatter, the EPFM-based fracture assessment of welded PP structures is a challenging task.

The application of fracture toughness values for failure assessment and dimensioning requires a reliable detection of crack-like defects in the component investigated and the determination of proper crack-tip loading parameters for various loading situations and at application temperatures. J integral values can be calculated for different crack configurations in a structure using finite element methods. These models use the numerical implementation of the original line integral definition of Rice [14, 32]. A proper material model is needed for these simulations, which considers the loading rate and temperature dependence of the elastic–plastic deformation behaviour of the specific polymer [32, 33].

The proper determination of J integral values for the dimensioning scheme above remains a challenging task to further investigations. As it was described previously, the critical J integral values can be determined experimentally by:

- Energy-based analysis (force-displacement curves) along with corresponding elastic and plastic geometry factors [9]

- Experimental analysis based on the original Rice J integral definition using such full-field strain analysis [34, 35]

The applicability to a proper dimensioning necessitates, however, both for the experimental and for the numerical values of the same physical basis and the same mathematical background. This, however, is a highly challenging task, and it requires the determination of experimental J values based on the original line integral definition of Rice [34]. The determination of fracture toughness and crack-tip loading must follow a consistent scheme and consider the inherent statistical character of both data.

Acknowledgements

The basic ideas and the first experiments regarding J integral and CTOD have been performed under the supervision of Prof R. W. Lang in Leoben; I would like to express our very great appreciation to him for his valuable support at that time. Our grateful thanks are also extended to Mrs. Theresa Distlbacher; her master's thesis supported the experiments described in this paper. I would also like to extend my thanks to the technicians, M. Lackner and G. Wurzinger, in my laboratory for supporting the actual experimental work.

Author details

Zoltan Major[1]*, Daniel Kimpfbeck[2,3] and Matei C. Miron[1]

1 Institute of Polymer Product Engineering, Johannes Kepler University Linz, Linz, Austria

2 Institute of Structural Lightweight Design, Johannes Kepler University Linz, Linz, Austria

3 Christian Doppler Laboratory for Structural Strength Control of Lightweight Constructions, Linz, Austria

*Address all correspondence to: zoltan.major@jku.at

References

[1] Blumenauer H, Pusch G. Technische Bruchmechanik. Leipzig Stuttgart: Deutscher Verlag für Grundstoffindustrie; 1993. 3 Auflage. ISBN: 3-342-00659-5; siehe AMK-Büchersammlung unter E 29-3

[2] Anderson TL. Fracture Mechanics. Fundamentals and Applications. 2nd ed. Boca Raton: CRC Press; 1995. 2. Auflage. ISBN: 978-0849342608; siehe AMK-Büchersammlung unter E 8–1

[3] Saxena A. Nonlinear Fracture Mechanics for Engineers. Boca Raton: CRC Press; 1997

[4] Williams JG. Fracture Mechanics of Polymers. In: Ellis Horwood Series in Engineering Science. Chichester: Horwood; 1984

[5] Kanninen MF, Popelar CH. Advanced Fracture Mechanics. New York: Oxford University Press; 1985

[6] Broek D. The Practical Use of Fracture Mechanics. Dordrecht: Kluwer Academic Publishers; 1989

[7] Wells AA. In: Proceedings Crack Propagation Symposium, Cranfield, England; 1961. Vol. 1. Paper B4

[8] Dugdale DS. Yielding of steel sheets containing slits. Journal of the Mechanics and Physics of Solids. 1960; **8**(2):100-104

[9] Joyce JA. Manual on Elastic-Plastic Fracture: Laboratory Test Procedures, ASTM Manual Series; MNL 27, ASTM Publication Code Number (PCN): 28–027096-30; 1996

[10] Xu S, Petri N, Tyson WR. Evaluation of CTOA from load vs. load-line displacement for C(T) specimen. Engineering Fracture Mechanics. 2009; **76**:2126-2134

[11] Samadian K. Measurement of CTOD along a surface crack by means of digital image correlation. Engineering Fracture Mechanics. 2019;**205**:470-485

[12] Zhu XK. Comparative study of CTOD-resistance curve test methods for SENT specimens. Engineering Fracture Mechanics. 2017;**172**:17-38

[13] Cherepanov GP. On crack propagation in continuous media. Applied Mechanics and Mathematics. 1967;**31**:476-488

[14] Rice JR. A path independent integral and the approximate analysis of strain concentration by notches and cracks. Journal of Applied Mechanics. 1968; **35**(2):379-386

[15] Hutchinson JW. Singular behaviour at the end of a tensile crack in a hardening material. Journal of the Mechanics and Physics of Solids. 1968; **16**:13-31

[16] Rice JR, Rosengren GF. Plane strain deformation near a crack tip in a power-law hardening material. Journal of the Mechanics and Physics of Solids. 1968; **16**:1-12

[17] Begley JA, Landes JD. The J-integral as a fracture criterion. In: National Symposium on Fracture Mechanics; 1971(part II, ASTM STP 514). pp. 1-20

[18] Zhu XK, Joyce JA. Review of fracture toughness (G, K, J, CTOD, CTOA) testing and standardization. Engineering Fracture Mechanics. 2012; **85**:1-46

[19] Zhu J. J-integral resistance curve testing and evaluation. Journal of Zhejiang University Science A. 2009; **10**(11):1541-1560

[20] Major Z. [PhD thesis]. A Fracture Mechanics Approach to Characterize

Rate Dependent Fracture Behavior of Engineering Polymers. University of Leoben; 2002

[21] Blackman B, Pavan A, Williams JG, editors. Fracture of Polymers Composites and Adhesives II, ESIS Publication 32. Kidlington, Oxford, UK: Elsevier; 2003

[22] Grellmann W, Seidler S, editors. Deformation and Fracture Behavior of Polymers. Berlin Heidelberg: Springer-Verlag; 2001

[23] Grellmann W, Che M. Assessment of temperature-dependent fracture behaviour with different fracture mechanics concepts on example of unoriented and cold-rolled polypropylene. Journal of Applied Polymer Science. 1997;**66**:1237-1249

[24] Grellmann W, Seidler S. (Hrsg.). Kunststoffprüfung. München: Carl Hanser Verlag; 2015. 3. Auflage, S. 258/259. ISBN: 978-3-446-44350-1; siehe AMK-Büchersammlung unter A 18)

[25] Frontini P. On the applicability of the load separation criterion to acrylonitrile/butadiene/styrene terpolymer resins. Polymer. 1996;**37**: 4033-4039

[26] Frontini PM. Non-linear fracture mechanics of polymers: Load separation and normalization methods. Engineering Fracture Mechanics. 2012; **79**:389-414

[27] Maspoch ML. Fracture characterization of ductile polymers through methods based on load separation. Polymer Testing. 2009;**28**: 204-208

[28] Baldi F. Application of the load separation criterion in J-testing of ductile polymers: A round-robin testing exercise. Polymer Testing. 2015;**44**: 72-81

[29] Distlbacher T. [MSc Thesis]. Experimental and numerical analysis of large-scale welded polymer structures. Johannes Kepler University; 2014

[30] Gosch A. [MSc thesis]. Examination of J-Integral methods and their applicability to tough polypropylene materials. University of Leoben; 2017

[31] Noels L. Fracture Mechanics, Damage and Fatigue Non Linear Fracture Mechanics: J-Integral, University of Liege, lecture Notes; 2014

[32] Bruyneel M. Computational Fracture Mechanics with SamCef. In: NAFEMS Nordic Regional Conference. Trends and Future Needs in Enigineering Simulation. 26–27 October 2010; Gotheburg; 2010

[33] Kimpfbeck D. [BSc Thesis]. Experimental characterisation of fracture mechanical behaviour in welded polymer interfaces. Johannes Kepler University; 2018

[34] Feichter C. [PhD Thesis]. Characterization of the fatigue behavior of elastomer compounds by fracture mechanics methods. University of Leoben; 2006

[35] Kolednik O. A new view on J-integrals in elastic–plastic materials. International Journal of Fracture. 2014; **187**:77-107

Chapter 3

Fatigue Limit Reliability Analysis for Notched Material with Some Kinds of Dense Inhomogeneities Using Fracture Mechanics

Tatsujiro Miyazaki, Shigeru Hamada and Hiroshi Noguchi

Abstract

This study proposes a quantitative method for predicting fatigue limit reliability of a notched metal containing inhomogeneities. Since the fatigue fracture origin of the notched metal cannot be determined in advance because of stress nonuniformity, randomly distributed particles, and scatter of a matrix, it is difficult to predict the fatigue limit. The present method utilizes a stress-strength model incorporating the "statistical hardness characteristics of a matrix under small indentation loads" and the "statistical hardness characteristics required for non-propagation of fatigue cracks from microstructural defects". The notch root is subdivided into small elements to eliminate the stress nonuniformity. The fatigue limit reliability is predicted by unifying the survival rates of the elements obtained by the stress-strength model according to the weakest link model. The method is applied to notched specimens of aluminum cast alloy JIS AC4B-T6 containing eutectic Si, Fe compounds and porosity. The fatigue strength reliability at 10^7 cycles, which corresponds to the fatigue limit reliability, is predicted. The fatigue limits of notch root radius ρ = 2, 1, 0.3, and 0.1 mm are obtained by rotating-bending fatigue tests. It is shown that the fatigue limits predicted by the present method are in good agreement with the experimental ones.

Keywords: metal fatigue, fatigue limit reliability, notch effect, aluminum cast alloy, inhomogeneity

1. Introduction

Aluminum cast alloys are widely applied, for example, in motor vehicles, ships, aircraft, machines, and structures, owing to the high cast ability and high specific strength [1–3]. They can be improved so as to meet specific mechanical properties by tuning the casting method, the alloying elements, and the cooling and heat treatment conditions [4–6]. Generally, precipitation hardening, also called age-hardening, is used to strengthen the aluminum cast alloys, which brings the dense precipitate of particles such as eutectic Si. The precipitations form fine microstructures such as dendrites, which significantly improve the mechanical properties.

However, the resultant stress concentrations by the precipitations further to fatigue fracture unfortunately [7–9]. Moreover, the possibility of the fatigue fracture increases more and more if microstructural flaws such as porosity are created in the casting process [10–15]. Because the precipitate particles and the microstructural defects are unique, the fatigue strength of the aluminum cast alloys is obliged to treat statistically.

Statistical fatigue test methods [16, 17] are standardized to determine the reliability of the fatigue strength. However, because they require many fatigue tests, it is time-consuming to determine the fatigue strength reliability at 10^7 stress cycles. Moreover, because the weakest region which controls the fatigue strength of the specimen is not known, the present materials cannot be improved rationally. Hence, a faster, rational method for quantitatively and nondestructively predicting the effect of inhomogeneities on fatigue strength is necessary for safe and reliable machine designs and for economical and quick material developments.

Several methods for predicting the fatigue strength at 10^7 stress cycles, which are equivalent to the statistically determined fatigue limit of aluminum cast alloys, have been proposed [18–24]. Through a series of stress analyses and fatigue experiments, Murakami et al. [18–20] clarified the non-propagation limit of a fatigue crack initiated by a microstructural defect and proposed a simple formula for predicting the fatigue limit of a plain specimen containing defects [18–20]. The non-propagation limit of a fatigue crack initiated by microstructural defect is determined by the defect size and mechanical characteristics of the matrix near the defect. The maximum defect, which is often estimated by extreme statistics, is therefore assumed to be the origin of the fatigue fracture. Most of the methods are based on the assumption that fatigue fracture begins at the maximum defects, and they often do not consider the interference effects of inhomogeneities and the scatter of the hardness of the matrix [25]. Because aluminum cast alloys have much higher densities of inhomogeneities, it is presumed that the interference effect is not negligible and the maximum inhomogeneity is not in the severest mechanical state necessarily. Additionally, in the case of a notched specimen, the stress varies significantly. The most severe mechanical defect should be used for prediction, even if it is not maximal. Generally, the fatigue limit of a notched specimen of a homogeneous metal in which microstructural defect is not the origin of the fatigue fracture consists of the microcrack and macrocrack non-propagation limits [26–32]. This fact is widely used in predicting fatigue limit. However, since microstructural defects act as crack initiation sites, the fatigue limit of an inhomogeneous metal also cannot be predicted by these two types of crack non-propagation limits.

In this study, a quantitative method for predicting the fatigue limit reliability of a notched metal containing inhomogeneous particles is proposed. The present method is also based on the stress-strength model and is applied to notched specimens of an Al-Si-Cu alloy (JIS AC4B). The inhomogeneous particle in the alloy comprises eutectic Si and Fe compounds and porosity in the matrix. Rotating-bending fatigue tests are performed on the notched specimens of AC4B-T6 by changing notch root radius variously. The validity of the present method is examined by comparing its numerical prediction with experimental results.

2. Crack non-propagation limits for predicting fatigue limit of notched specimen

Generally, when fatigue tests are performed on a notched specimen by changing the notch root radius ρ for a given notch depth t, the typical relationship between the fatigue limit σ_w and ρ is as shown in **Figure 1**; here, σ_{w0} is the fatigue limit of the

Nomenclature	
t	notch depth
ρ	notch root radius
ρ_0	branch point
ρ_d	limit notch root radius
σ_w	fatigue limit of notched specimen
σ_{w0}	fatigue limit of plain specimen
σ_{w1}	microcrack non-propagation limit
σ_{w2}	long macrocrack non-propagation limit
σ_{wd}	small macrocrack non-propagation limit

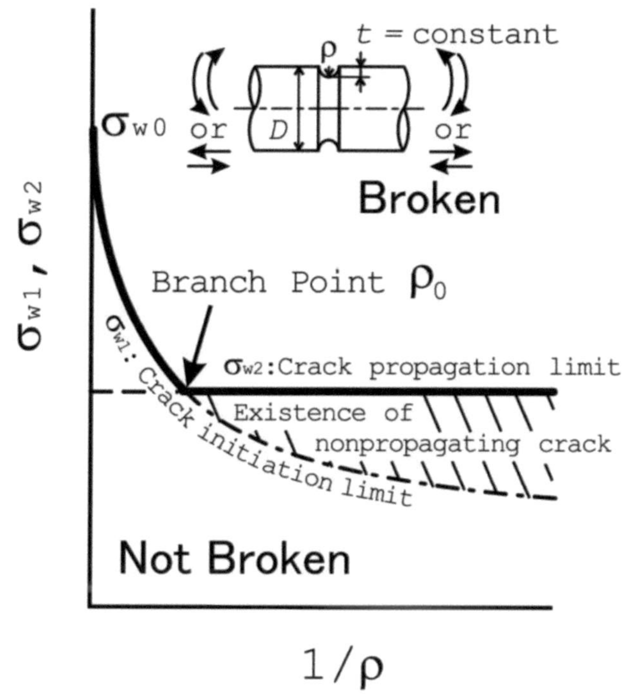

Figure 1.
Schematic illustration of fatigue limit of a notched structure without defects.

plain specimen, σ_{w1} is the microcrack non-propagation limit, σ_{w2} is the macrocrack non-propagation limit, and ρ_0 is a material property known as the branch point, the critical value of which determines whether the non-propagating crack exists along the notch root [26, 27]. If the notch is sufficiently deep, ρ_0 is constant [27].

If $\rho > \rho_0$, σ_{w1} is the fatigue limit [33]. σ_{w1} can be predicted from the mechanical characteristics of the microstructure. Conversely, if $\rho \le \rho_0$, σ_{w2} is the fatigue limit [33]. σ_{w2} is constant and independent of ρ. This means that the σ_{w2} is equal to the fatigue limit of the cracked specimen as $\rho \to 0$. That is, the notch can be assumed to be a crack and σ_{w2} can be predicted by the fracture mechanics.

In the case of metals containing microstructural defects, the non-propagation limit of the fatigue crack that originates from the microstructural defect may be the fatigue limit. Because the defect is categorized as a macrocrack, the low macrocrack non-propagation limit is differentiated from σ_{w2}. The threshold stress intensity factor range ΔK_{th} determines whether the fatigue crack originating from the macrocrack is arrested. The value of ΔK_{th} is an indication of the dependency of the different crack lengths [20]. In this study, a crack for which ΔK_{th} is constant irrespective of its length, and which exhibits the small-scale yielding (SSY), is defined as a long macrocrack. Conversely, a crack for which ΔK_{th} is dependent on the length, and which exhibits the large-scale yielding (LSY), is defined as a small macrocrack [33, 34]. The three following types of crack non-propagation limits are introduced and defined to predict the fatigue limit of a notched specimen of aluminum cast alloy [35]:

σ_{w1}:	This is the non-propagation limit of a microcrack that is initiated by repeated irreversible plastic strains in a homogeneous notch stress field without microstructural and structural stress concentrations
σ_{wd}:	This is the non-propagation limit of a three-dimensional fatigue crack that originates from microstructural defects such as nonmetallic inclusions and pits in a homogeneous notch stress field without other microstructural and structural stress concentrations
σ_{w2}:	This is the non-propagation limit of structural long macrocracks such as deep notches with $\rho < \rho_0$

Figure 2 is a schematic illustration of the relationships between ρ and each of σ_{w1}, σ_{wd}, and σ_{w2}. Further, ρ_0 and ρ_d are, respectively, the branch point and limit notch root radius, which determines whether the fatigue limit is affected by the microstructural defects. σ_{w1} and σ_{wd} decrease as ρ decreases, whereas σ_{w2} attains a constant value and becomes independent of ρ. If $\rho \geq \rho_d$, σ_{wd} is equal to the fatigue limit σ_w. If $\rho_0 < \rho < \rho_d$, σ_{w1} is equal to σ_w. If $\rho \leq \rho_0$, σ_{w1} and σ_{wd} are cut off by σ_{w2}, and σ_{w2} is equal to σ_w.

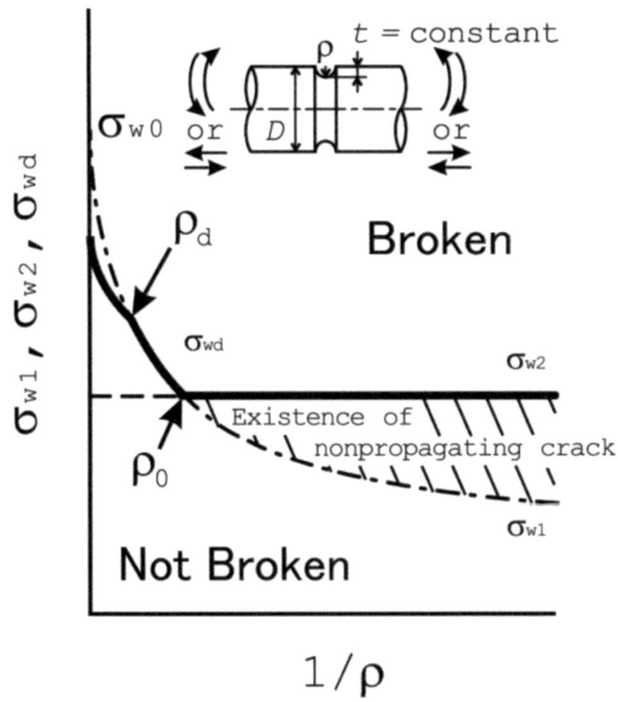

Figure 2.
Schematic illustration of fatigue limit of a notched structure with defects.

Because the hardness is locally scattered and numerous defects are distributed through the material, the microcrack and defect that determine the fatigue fracture cannot be determined in advance. In this situation, the probabilities of the arrest of the microcrack and the fatigue crack originating from the defect are, respectively, determined by the statistical characteristics of the hardness and the statistical characteristics of the defect. That is, σ_{w1} and σ_{wd} are, respectively, described by probability distributions.

3. Method for predicting fatigue limit reliability of notched metal containing inhomogeneous particles

Nomenclature	
A_j	size of jth surface element
A_j^*	size of jth region where the relative first principal stress corresponds to $\sigma_{1,j}^*$
A_{npc}, $A_{npc\ R}$, $A_{npc\ P}$	region required for the non-propagation of fatigue crack
$\sqrt{area_P}$	size of surface defect
$\sqrt{area_{P_1}}$	lower limit size of small surface crack
$\sqrt{area_R}$	size of internal defect
F, F_P, F_R	geometric correction factor
F_{σ_w}	fatigue limit reliability of notched specimen
$f_{H_{VM}1}$, $f_{H_{VM}S}$, $f_{H_{VM}R}$, $f_{H_{VM}P}$	H_{VM} distribution
f_{χ^2}	χ^2 distribution
g_P, g_R, g_S	limit hardness
γ_m	stress relaxation effect
H_V, H_{VM}	Vickers hardness
K_{In}, K_{IP}, K_{IR}	stress intensity factor
K_t	stress concentration factor
ΔK_w	threshold stress intensity factor range
ΔK_{wLL}	lower limit value of ΔK_w
ΔK_{wUL}	upper limit value of ΔK_w
$M_{S0}\left(\sqrt{area_{P_0}}\right)$	the number of surface cracks with $\sqrt{area_P} \geq \sqrt{area_{P_0}}$ in a unit area
$M_{V0}(R_0)$	the number of particles with $R \geq R_0$ in a unit volume
M_d	types of inhomogeneous particles
$\overline{N}_{V0}$	the number of particles in a unit volume
n_S	the number of surface elements
n_V	the number of solid elements
P, P_R	indentation load
$P_V(R_0)$	existence probability of particles with $R \geq R_0$
R_c	limit size of small interior crack
S_{σ_w}	survival rate of notched specimen
$S_{\sigma_{w1}}$	survival rate of surface element with microcracks

$S_{\sigma_{wd}}, S_{\sigma_{wdI}}, S_{\sigma_{wdS}}$	survival rate of solid element with microstructural defects
$s^2_{H_{VM}}$	population of H_{VM} distribution
V_j	size of jth solid element
λ, ν	material constant
$\mu_{H_{VM}}$	mean of H_{VM} distribution
$\sigma_1, \sigma_{1,j}$	first principal stress
$\sigma_1^*, \sigma_{1,j}^*$	relative first principal stress
$\sigma_{ym}(T_m, Z_m)$, $\sigma_{zm}(T_m, Z_m)$	stress produced by the spherical particle in the infinite body under $\sigma_z = Z_m$ and $\sigma_x = \sigma_y = \sigma_z = T_m$
σ_m	mean stress
σ_n	stress amplitude
χ_{σ_1}	stress gradient of first principal stress

This section presents a method for predicting the fatigue limit reliability of a notched specimen with stress concentration factor K_t, notch depth t, and notch root radius ρ under zero mean stress. The control volume is actually divided into surface and solid elements so that the stresses applied to the elements can be assumed to be constant. The fatigue strengths of all the elements are then stochastically evaluated by the stress-strength model on the mesoscale. The fatigue limit reliability is also predicted by assembling the fatigue strengths using the weakest link model [25].

3.1 Stress relaxation effect of interference of inhomogeneous particles

Figure 3 is a schematic illustration of the analytical model of a metal containing inhomogeneous particles. The metal is approximated by a cubic lattice model to determine the stress relaxation effect of the interference of the particles [36].

3.1.1 Statistical characteristics of inhomogeneous particles

The probability of existence of such particles is given by the following equation [37]:

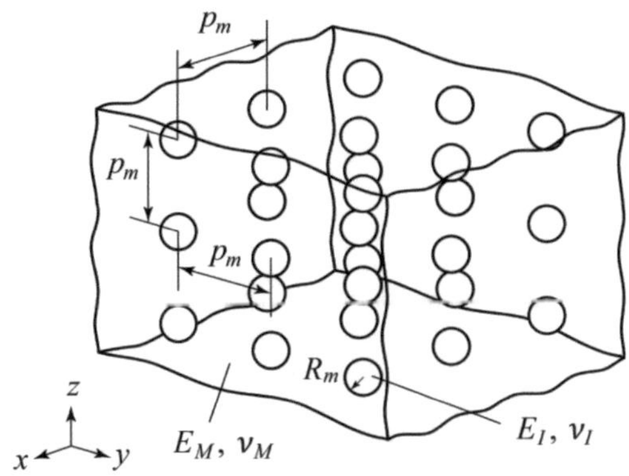

Figure 3.
Approximate model of metal with inhomogeneous particles.

$$P_V(R_0) = \exp\left\{-\left(\frac{R_0}{\lambda}\right)^{\nu}\right\}. \tag{1}$$

Here, ν and λ are material constants, R_0 is the particle radius, and $P_V(R_0)$ is the probability of the existence of particles with radii greater than R_0.

The total number of particles in a unit volume is denoted by $\overline{N}_{V0}$. The average number of particles with radii greater than R_0 in a unit volume, $M_{V0}(R_0)$, is given by the following equation [37]:

$$M_{V0}(R_0) = \overline{N}_{V0} \cdot P_V(R_0). \tag{2}$$

A particle cross-sectioned by the specimen surface is projected onto a plane perpendicular to the first principal stress. The projected area is then modified as shown in **Figure 4** by considering the mechanics. The modified area is denoted by $area_P$. The average number of cross-sectioned particles with areas larger than $\sqrt{area_{P_0}}$ in a unit area, $M_{S0}\left(\sqrt{area_{P_0}}\right)$, is given by the following equation [12, 24]:

$$M_{S0}\left(\sqrt{area_{P_0}}\right) = \lambda\,\overline{N}_{V0}\int_0^1 \frac{t}{\sqrt{1-t^2}}\left\{\Gamma\left(1+\frac{1}{\nu},\left(\frac{\sqrt{area_{P_0}}}{\lambda\,\theta^-}\right)^{\nu}\right)\right.$$

$$\left.+\Gamma\left(1+\frac{1}{\nu},\left(\frac{\sqrt{area_{P_0}}}{\lambda\,\theta^+}\right)^{\nu}\right)\right\}dt, \tag{3}$$

$$\theta^+ = \frac{\pi}{2} + 2\sqrt{1-t^?}, \tag{4}$$

$$\theta^- = \sin^{-1}t - t\sqrt{1-t^2}. \tag{5}$$

Here, Γ is a gamma function of the second kind.

3.1.2 Average radius and distance

The average particle radius is evaluated by the following equation:

$$R_m = \int_0^{\infty} R\frac{dP_V}{R}dR = \lambda\,\Gamma\left(1+\frac{1}{\nu}\right). \tag{6}$$

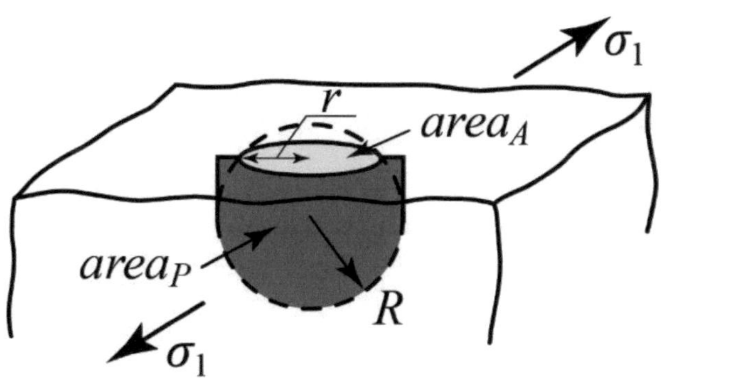

Figure 4.
Spheroidal particle cut by surface.

If $\overline{N}_{V0}$ particles are regularly arranged in a unit volume as shown in **Figure 3**, the average distance between the particles is evaluated by the following equation:

$$p_m = \left(\frac{1}{\overline{N}_{V0}}\right)^{1/3} \tag{7}$$

3.1.3 Stress relaxation effect

Nisitani [38] proposed a method for approximately solving the interference problem of notches by superposing simple basic solutions to satisfy the equilibrium conditions at the stress concentration point.

When the uniform tensile stress at infinity, $\sigma_{z\infty} = 1$, is applied to an infinite body, it is supposed that a stress field composed of $\sigma_z = Z_m$ and $\sigma_x = \sigma_y = \sigma_z = T_m$ is formed around the particle. T_m and Z_m are set to satisfy the equilibrium condition at point (0, 0, R_m). Because the stress acting on a single particle in the z-direction, $T_m + Z_m$, is composed of $\sigma_{z\infty} = 1$ and the stresses due to the other particles, the stress equilibrium condition in the z-direction is as follows:

$$T_m + Z_m = 1 + \sum \sum_{\substack{(i,j,k) \neq (0,0,0) \\ i,j,k=-\infty}} {}^{\infty} \sum \sigma_{zm}(T_m, Z_m)\Big|_{\substack{x_{i,j,k} = -ip_m \\ y_{i,j,k} = -jp_m \\ z_{i,j,k} = R_m - kp_m}} . \tag{8}$$

Here, $\sigma_{zm}(T_m, Z_m)$ is the stress in the z-direction at (0, 0, R_m) produced by the spherical particle located at (ip_m, jp_m, kp_m) in the infinite body under $\sigma_z = Z_m$ and $\sigma_x = \sigma_y = \sigma_z = T_m$.

The stress equilibrium condition in the y-direction is also given by

$$T_m = \sum \sum_{\substack{(i,j,k) \neq (0,0,0) \\ i,j,k=-\infty}} {}^{\infty} \sum \sigma_{ym}(T_m, Z_m)\Big|_{\substack{x_{i,j,k} = -ip_m \\ y_{i,j,k} = -jp_m \\ z_{i,j,k} = R_m - kp_m}} . \tag{9}$$

Here, $\sigma_{ym}(T_m, Z_m)$ is the stress in the y-direction at (0, 0, R_m) produced by the spherical particle located at (ip_m, jp_m, kp_m) in the infinite body under $\sigma_z = Z_m$ and $\sigma_x = \sigma_y = \sigma_z = T_m$.

T_m and Z_m are obtained by solving the simultaneous linear Eqs. (8) and (9). In this study, the stress relaxation effect γ_m of the interference of the particles is assumed to be

$$\gamma_m = T_m + Z_m. \tag{10}$$

3.1.4 Characteristics of elastic stress field near notch root

If the notch is sufficiently deep, a unique stress field determined by the maximum stress and ρ is formed near the notch root [27]. The first principal stress normalized by the maximum stress at the notch root is denoted by σ_1^*. **Figure 5** shows a contour map of $\sigma_1^* = [0.4, 1]$ near the notch root in a semi-infinite plate under tensile stress [35]. The value of σ_1^* is independent of the notch shape [27]. If the notch root is divided as shown in **Figure 5**, the length of the notch edge and size of j-th region in which the relative first principal stress is $\sigma_{1,j}^*$ are denoted by l_j^* and A_j^* $(j = 1, \cdots)$, respectively. The values of l_j^* and A_j^* are given in **Table 1** [35].

To predict the fatigue limit reliability, the control volume is set at the notch root and divided into surface and solid elements. The sizes of the solid and surface

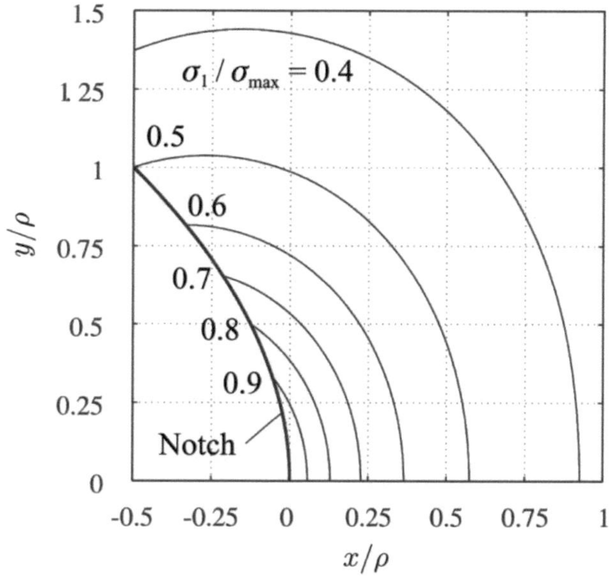

Figure 5.
Contour map of relative first principal stress $\sigma_1^* = [0.4, 1]$ *near the notch root.*

j	1	2	3	4	5	6
$\sigma_{1,j}^*$	1–0.95	0.95–0.9	0.9–0.8	0.8–0.7	0.7–0.6	0.6–0.5
l_j^*/ρ	0.463	0.216	0.361	0.358	0.402	0.496
A_j^*/ρ	0.0083	0.0181	0.0703	0.144	0.296	0.657

Table 1.
Area of isostress near the notch root.

elements are denoted by V_j and A_j $(j = 1, \cdots)$, respectively. In the case of a typical notch, as in the notched specimen, the control volume can be divided as shown in **Figure 5**. For example, when a circular bar with a circumferential notch is divided, V_j and A_j are approximated as follows [35]:

$$V_j \cong A_j^* \times (\text{average diameter of } j \text{ th region}) \times \pi, \tag{11}$$

$$A_j \cong l_j^* \times (\text{average diameter of } j \text{ th region}) \times \pi. \tag{12}$$

3.2 Fatigue survival rate of surface element containing microcracks

3.2.1 Statistical characteristics of Vickers hardness

The authors proposed a virtual small cell model for predicting the statistical characteristics of the Vickers hardness in a small region [25, 35]. If the population of the virtual small cells is described by an arbitrary distribution of the mean μ and variance s^2, the statistical characteristics of the Vickers hardness can be described by the normal distribution of the mean μ and the variance s^2/n_c, based on the central limit theory, where n_c is the number of virtual small cells in the indentation area.

If m_1 Vickers hardness values are measured in this way using an indentation load P, their statistical characteristics are described by the following normal distribution [25, 35]:

$$f_{H_{VM}1}(H_{VM}) = \frac{1}{\sqrt{2\pi s^2_{H_{VM}1}}} \exp\left\{-\frac{\left(H_{VM} - \mu_{H_{VM}1}\right)^2}{2s^2_{H_{VM}1}}\right\}. \tag{13}$$

Here, H_{VM} is the Vickers hardness of the matrix that does not contain microstructural defects, $\mu_{H_{VM}1}$ is the sample mean, and $s^2_{H_{VM}1}$ is the sample variance.

Based on the central limit theorem, the relationship between the sample mean $\mu_{H_{VM}1}$ and the population mean $\mu_{H_{VM}0}$ is $\mu_{H_{VM}1} = \mu_{H_{VM}0}$. Further, the relationship between the sample variance $s^2_{H_{VM}1}$ and the population variance $s^2_{H_{VM}0}$ is described by the χ^2 distribution with the freedom degree of $n = m_1 - 1$ [25, 35]:

$$f_{\chi^2}(\chi^2) = \frac{1}{2\Gamma(n/2)}\left(\frac{\chi^2}{2}\right)^{\frac{n}{2}-1} \exp\left(-\frac{\chi^2}{2}\right) \tag{14}$$

$$\Gamma(t) = \int_0^\infty x^{t-1}e^{-x}dx, \chi^2 = m_1\frac{s^2_{H_{VM}1}}{s^2_{H_{VM}0}} \tag{15}$$

3.2.2 Fatigue survival rate of surface element containing microcracks

The microcrack non-propagation limit σ_{w0} is determined by the average characteristics of the material properties around the microcrack. σ_{w0} can be empirically predicted by the following equations [20, 26]:

$$\sigma_{w0}\big|_{\sigma_m=0} = 1.6\, H_{VM}, \tag{16}$$

$$\sigma_{w1}\big|_{\sigma_m=0} = \frac{\sigma_{w0}\big|_{\sigma_m=0}}{K_t\sqrt{1+4.5\varepsilon_0\big|_{\sigma_m=0}/\rho}} = f(H_{VM}, \rho, \sigma_m = 0, K_t). \tag{17}$$

(σ_{w0}, σ_{w1}, and σ_m are in MPa, H_{VM} is in kgf/mm^2, and ρ *and* ε_0 are in mm.)

If the stress relaxation effect γ_m is considered, $\gamma_m\ \sigma_{1,j}$ is applied to j-th surface element. Because fatigue fracture occurs when $\gamma_m\ \sigma_{1,j}/K_t$ is greater than σ_{w1}, the limit hardness $g_S(\sigma_{1,j})$ that determines the occurrence is given by the following equation:

$$g_S = f^{-1}(H_{VM}, \rho, \sigma_m = 0, K_t)\big|_{\sigma_{w1}=\gamma_m\ \sigma_{1,j}/K_t}. \tag{18}$$

Here, f^{-1} is a function obtained by solving Eq. (18) on H_{VM}.

It is supposed that $A_{npc\ S}$ exhibits the microcrack non-propagation limit characteristics in Eqs. (16) and (17). If H_{VM} of the matrix is greater than g_S in $A_{npc\ S}$, fatigue fracture will not occur in $A_{npc\ S}$. Therefore, the probability $S_{\sigma_{w1}0,j}$ that fatigue fracture does not occur in $A_{npc\ S}$ below $\sigma_{1,j}$ is given by [25, 35]

$$S_{\sigma_{w1}0,j} = \int_{g_S}^\infty f_{H_{VM}S}(h_{vm})dh_{vm}. \tag{19}$$

Here, $f_{H_{VM}S}$ is the normal distribution of $\mu_{H_{VM}S}$ and $s^2_{H_{VM}S}$.

The fatigue survival rate $S_{\sigma_{w1},j}$ of jth surface element containing microcracks is given by the following equation [25, 35]:

$$S_{\sigma_{w1},j} = \int_0^\infty f_{\chi^2} \cdot \left(S_{\sigma_{w1}0,j}\right)^{\frac{A_j}{A_{npc\ S}}} d\chi^2. \tag{20}$$

If fatigue fracture does not occur in all the surface elements, the notched specimen would not be broken by the microcrack. Therefore, the fatigue survival rate of a surface element containing microcracks, $S_{\sigma_{w1}}$, is obtained by multiplying the fatigue survival rates of all the surface elements as follows [25, 35]:

$$S_{\sigma_{w1}} = \prod_{j=1}^{n_S} S_{\sigma_{w1},j}. \tag{21}$$

Here, n_S is the number of surface elements.

3.3 Fatigue survival rate of surface element containing microstructural defects

The authors [25] proposed a method for predicting the reliability of the small macrocrack non-propagation limit for a nonzero stress gradient using the "statistical hardness characteristics of a matrix under small indentation loads" and the "statistical hardness characteristics required for non-propagation of fatigue cracks originating from microstructural defects in a material" [25]. The stress relaxation effect was introduced into the method to make it applicable to a metal containing dense inhomogeneous particles.

σ_{wd} is divided into two crack non-propagation limits, namely, the non-propagation limit σ_{wdI} of the small crack originating from the interior defect and the non-propagation limit σ_{wdS} of the small crack originating from the surface defect.

3.3.1 Fatigue survival rate of solid element containing interior microstructural defects

Because the fatigue crack that originates from a defect propagates on the plane perpendicular to the first principal radial stress, a spherical particle of radius R is projected onto this plane and assumed to be a penny-shaped crack. If the projected area is denoted by $area_R$, its square root is related to R as follows:

$$\sqrt{area_R} = \sqrt{\pi}R. \tag{22}$$

The stress intensity factor K_{IR} of the small interior crack is given by [20, 35]

$$K_{IR} = 0.5\, F_R\, \gamma_m\, \sigma_{1,j}\sqrt{\pi\sqrt{area_R}}, \tag{23}$$

$$F_R = \frac{4}{\pi^{5/4}}\left\{1 - \left(\frac{2}{\sqrt{\pi}} - \frac{4}{3\,\pi}\right)\frac{\sqrt{area_R}}{\rho}\right\}. \tag{24}$$

Moreover, the threshold stress intensity factor range ΔK_w of the small surface defect of size $\sqrt{area_P}$ in the metal with Vickers hardness H_{VM} is given by the following equation [33, 34]:

$$\Delta K_w\big|_{\sigma_m=0} = \frac{2\alpha\beta\sqrt{area_P}^{1/3}}{\ln\left(2\beta/H_{VM}+1\right)}, \tag{25}$$

$\alpha = 3.3 \times 10^{-3}$and $\beta = 120$.

(ΔK_w is in MPa$\sqrt{\text{m}}$, H_{VM} is in kgf/mm^2, and $\sqrt{area_P}$ is in μm).

The limit hardness that determines whether the fatigue crack originating from the interior microstructural crack is arrested, $g_R\left(\sigma_{1,j}, \sqrt{area_R}\right)$, is given based on the relationship $\sqrt{area_R} = 1.7\sqrt{area_P}$ by the following equation [25, 35]:

$$g_R = 240/\left\{\exp\left(\frac{1.56 \times 240}{F_R \gamma_m\ \sigma_{1,j} \sqrt{area_R}^{1/6}}\right) - 1\right\} \tag{26}$$

($\sigma_{1,j}$ is in MPa, g_R is in kgf/mm^2, and $\sqrt{area_R}$ is in μm.)

The relationship between $\mu_{H_{VM}R}$ and $\mu_{H_{VM}0}$ can be expressed as follows [25, 35]:

$$\mu_{H_{VM}R} = \mu_{H_{VM}0} = \mu_{H_{VM}1}. \tag{27}$$

Moreover, the relationship between $s^2_{H_{VM}R}$ and $s^2_{H_{VM}0}$ can be expressed as follows [25, 35]:

$$A_{npc\ R} \cdot s^2_{H_{VM}R} = A_{H_{VM}0} \cdot s^2_{H_{VM}0}. \tag{28}$$

Here, $A_{H_{VM}0} = P/\mu_{H_{VM}0}$, $A_{npc\ R} = P_R/g_R$, and P_R are the loads used to create the indentation for obtaining the Vickers hardness g_R and the indentation area $A_{npc\ R}$.

The fatigue survival rate of j-th solid element containing interior defects, $S^{(m)}_{\sigma_{wdI},j}$, is given by the following equation [25, 35]:

$$S^{(m)}_{\sigma_{wdI},j} = \int_0^{\infty} f_{\chi^2} \exp\left\{-\int_{\infty}^{R_c}\left(\int_0^{g_R} p^{(m)}_{dI} \frac{dM^{(m)}_{V,j}}{dR} f_{H_{VM}R} dh_{vm}\right) dR\right\} d\chi^2. \tag{29}$$

If the fatigue fracture does not occur in all the solid elements, the notched specimen would not be broken by the small interior defect. Therefore, the fatigue survival rate $S^{(m)}_{\sigma_{wdI}}$ of a solid element containing interior microstructural defects is obtained by multiplying the fatigue survival rates of all the solid elements as follows [25, 35]:

$$S^{(m)}_{\sigma_{wdI}} = \prod_{j=1}^{n_V} S^{(m)}_{\sigma_{wdI},j}. \tag{30}$$

Here, n_V is the number of solid elements.

3.3.2 Fatigue survival rate of surface element containing surface microstructural cracks

The stress intensity factor K_{IP} of a small surface crack of size $\sqrt{area_P}$ is given by the following equation [20, 35]:

$$K_{IP} = 0.65\, F_R\, \gamma_m\ \sigma_{1,j} \sqrt{\pi \sqrt{area_P}}, \tag{31}$$

$$F_P = 0.968 - 1.021 \frac{\sqrt{area_P}}{\rho}. \tag{32}$$

Further, the limit hardness $g_P(\sigma_{1,j}, \sqrt{area_P})$ that determines whether the small surface crack is arrested is given by the following equation [25, 35]:

$$g_P = 240/\left\{\exp\left(\frac{1.43 \times 240}{F_R\, \gamma_m\ \sigma_{1,j} \sqrt{area_P}^{1/6}}\right) - 1\right\}. \tag{33}$$

($\sigma_{1,j}$ is in MPa, g_P is in kgf/mm^2, and $\sqrt{area_P}$ is in μm.)

The fatigue survival rate of j-th surface element containing surface microstructural cracks, $S^{(m)}_{\sigma_{wdS},j}$, is given by

$$S^{(m)}_{\sigma_{wdS},j} = \int_0^{\infty} f_{\chi^2} \exp\left\{-\int_{\infty}^{\sqrt{area_{P_c}}}\left(\int_0^{g_P} p^{(m)}_{dS} \frac{dM^{(m)}_{S,j}}{d\sqrt{area_P}} f_{H_{VM}P} dh_{vm}\right) d\sqrt{area_P}\right\} d\chi^2 \quad (34)$$

The fatigue survival rate $S^{(m)}_{\sigma_{wdS}}$ of a surface element containing surface microstructural cracks is obtained by multiplying the fatigue survival rates of all the surface elements as follows [25, 35]:

$$S^{(m)}_{\sigma_{wdS}} = \prod_{j=1}^{n_S} S^{(m)}_{\sigma_{wdS},j}. \quad (35)$$

3.3.3 Fatigue survival rate of element with microstructural defects

The fatigue survival rate $S^{(m)}_{\sigma_{wd}}$ of an element containing microstructural defects is obtained by multiplying $S^{(m)}_{\sigma_{wdI}}$ and $S^{(m)}_{\sigma_{wdS}}$ as follows [25, 35]:

$$S^{(m)}_{\sigma_{wd}} = S^{(m)}_{\sigma_{wdI}} \times S^{(m)}_{\sigma_{wdS}}. \quad (36)$$

Because the material contains M_d types of inhomogeneous particles, the fatigue survival rate $S_{\sigma_{wd}}$ is given by the following equation:

$$S_{\sigma_{wd}} = \prod_{m=1}^{M_d} S^{(m)}_{\sigma_{wd}}. \quad (37)$$

3.4 Prediction of long macrocrack non-propagation limit σ_{w2}

σ_{w2} of the notched specimen with $\rho \leq \rho_0$ is equal to the fatigue limit of the cracked specimen obtained by $\rho \to 0$. ΔK_{wUL} is the upper limit of ΔK_w and is constant regardless of the crack length. $\sigma_{w2}|_{\sigma_m=0}$ can be obtained as follows [35]:

$$\sigma_{w2}|_{\sigma_m=0} = \frac{\Delta K_{wUL}|_{\sigma_m=0}}{2F\sqrt{\pi t}}. \quad (38)$$

3.5 Prediction of fatigue limit reliability

The probability that fatigue fracture is caused by microcracks or microstructural defects is obtained by the complementary event defined by the product of $S_{\sigma_{w1}}$ and $S_{\sigma_{wd}}$. Because the fatigue limit of a notched specimen, σ_w cannot be lower than σ_{w2}, and its fatigue limit reliability F_{σ_w} is obtained as follows [35]:

$$F_{\sigma_w} = \begin{cases} 0 & (\sigma_w \leq \sigma_{w2}) \\ 1 - S_{\sigma_{w1}} \times S_{\sigma_{wd}} & (\sigma_w > \sigma_{w2}) \end{cases} \quad (39)$$

4. Fatigue experiment

4.1 Material, shape of specimen, and experimental procedure

The material used for the experiment was Al-Si-Cu alloy (JIS AC4B). The age-hardened aluminum cast alloy is identified as AC4B-T6. **Table 2** shows its mechanical properties.

E	$\sigma_{0.2}$	σ_B	δ	H_V	H_{VM}
				9.8 N, 30 sec	29.4 mN, 30 sec
74	292	349	1.5	152	92
E: Young's modulus (GPa)				$\sigma_{0.2}$: 0.2% proof stress (MPa)	
σ_B: ultimate tensile strength (MPa)				δ: elongation (%)	
H_V: Vickers hardness of matrix with inhomogeneous particles (kgf/mm^2)					
H_{VM}: Vickers hardness of matrix without inhomogeneous particles (kgf/mm^2)					

Table 2.
Mechanical properties.

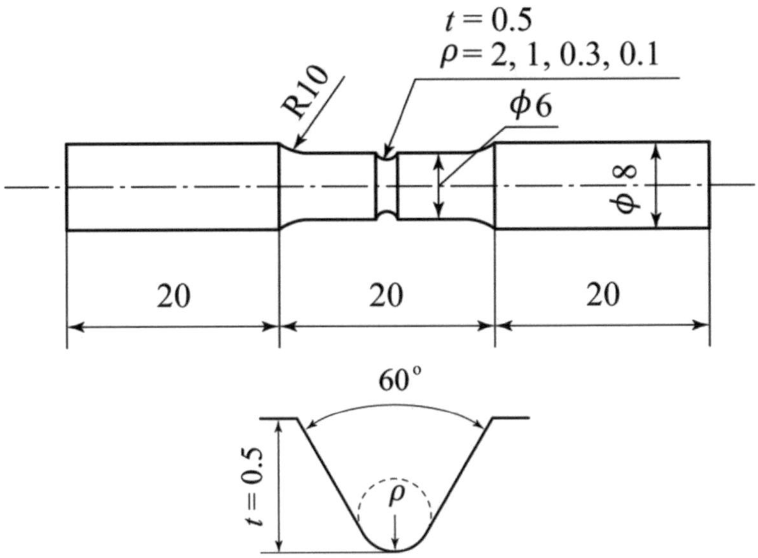

Figure 6.
Specimen configuration.

Figure 6 shows the configurations of the specimens. The notch depth t and opening angle θ were set at 0.5 mm and 60°, respectively. The notch root radii ρ were set at 2, 1, 0.3, and 0.1 mm, respectively. All the specimens were machined; polished with fine emery paper, alumina (3 μm), and diamond paste (1 μm); and also chemically polished. Rotating-bending fatigue tests were carried out according to JIS Z2274 and the earlier studies [27–31] under stress amplitude $\sigma_a = 60$–110 MPa and frequency $f = 50$ Hz. An Ono-type rotating-bending fatigue machine of a capacity 15 Nm was used for the tests. The nominal stress σ_n used for the analyses of the experimental results was the stress at the minimum cross section, where the diameter d was 5 mm. The fatigue life N_f was defined as the total number of stress cycles to failure.

4.2 Experimental results

Figure 7 shows S-N curves obtained from the results of the tests. **Figure 8** shows optical micrographs of a specimen when ρ = 0.1 mm for σ_w = 90 MPa. Fatigue limit $\sigma_w = 105$ when $\rho = 2$ mm. σ_w = 95 when $\rho = 1$ and 0.3 mm. σ_w = 90 when ρ = 0.1 mm. Since the non-propagating macrocrack was not observed at the fatigue

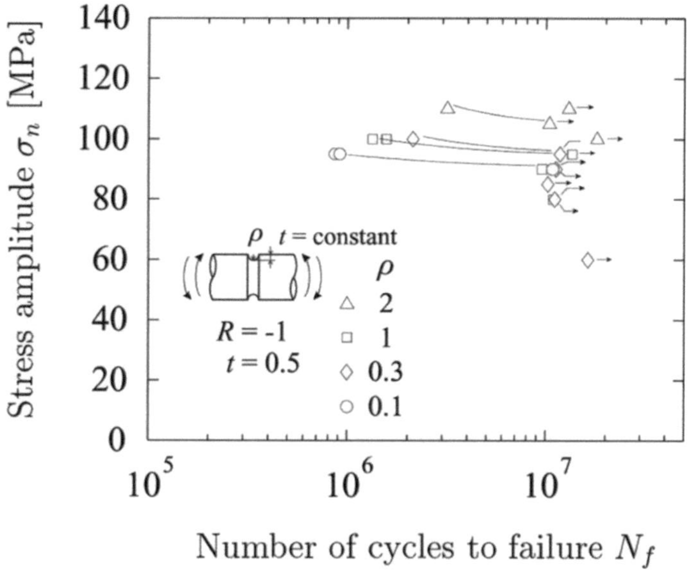

Figure 7.
S − N curve.

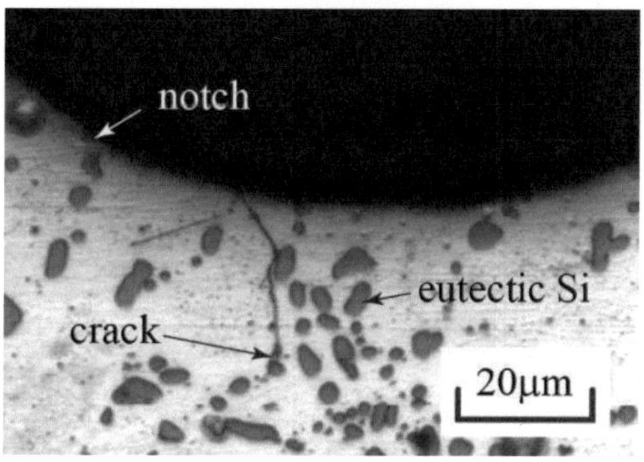

Figure 8.
Optical micrograph of non-propagating crack under notch root.

limit when $\rho = 1$ and 2 mm, it can be said that the microcrack non-propagating limit σ_{w1} or the small macrocrack non-propagating limit σ_{wd} appears as the fatigue limit. When ρ = 0.1 mm for σ_w = 90 MPa and ρ = 0.3 mm for σ_w = 95 MPa, the non-propagating macrocracks were observed along the notch root. Therefore, the long macrocrack non-propagating limit σ_{w2} = 90 MPa.

5. Examination of validity of prediction method

5.1 Notch sensitivity to crack initiation limit in age-hardened aluminum alloy

Figure 9 shows the relationship between $K_t\sigma_{w1}/\sigma_{w0}$ and ρ using early fatigue data of previous studies [28–30]. ε_0 values of the curves are shown in **Figure 10**. When ε_0 values were approximated with the lines by the least squares method, the following equation was obtained:

$$\varepsilon_0|_{\sigma_m=0} = 5.0 \times 10^{-4} H_B - 0.0164, \quad (40)$$
$$\rho \geq 0.5, 97 \leq H_B \leq 207.$$

(ε_0 and ρ are in mm, and H_B is in kgf/mm^2.)
Once σ_{w1} has been predicted, H_{VM} can be used instead of H_B.

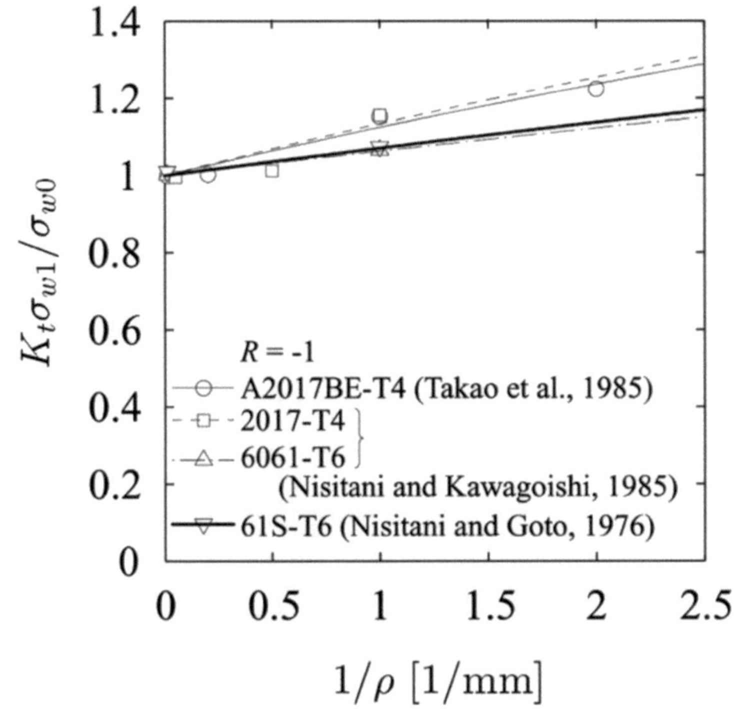

Figure 9.
Relation between $K_t\sigma_{w1}/\sigma_{w0}$ and $1/\rho$.

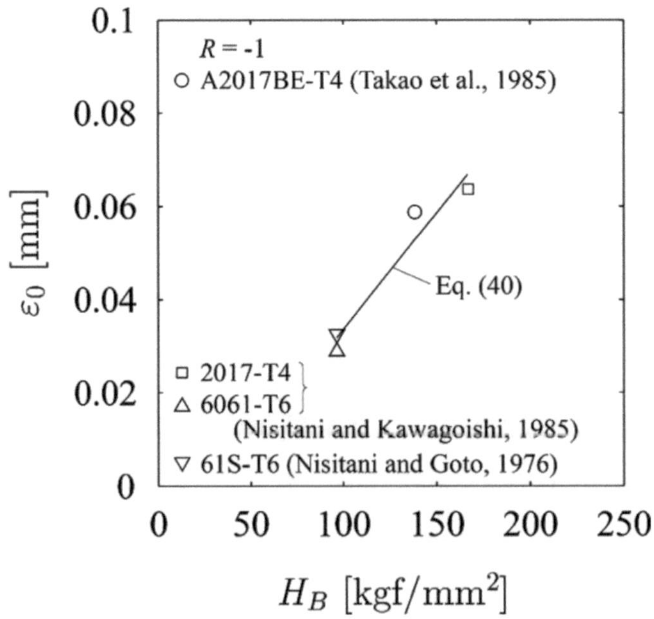

Figure 10.
Relation between ε_0 and H_B.

5.2 ΔK_{wUL} of age-hardened aluminum alloy

Figure 11 shows the values of ΔK_{wUL} obtained from the early fatigue data of σ_{w2} for different Al-Si-X alloys, where X is a transition element [28, 29, 31]. An approximation of ΔK_{wUL} obtained from the lines by the least squares method was used to derive the following equation:

$$\Delta K_{wUL}\big|_{\sigma_m=0} = \Delta K_{wLL} + 0.03\,H_B, \qquad (41)$$
$$40 \leq H_B \leq 100.$$

(ΔK_{wUL} and ΔK_{wLL} are in MPa$\sqrt{\text{m}}$, and $H_B$is in kgf/mm^2.)

Here, ΔK_{wLL} is the lower limit of ΔK_w and ΔK_{wLL} = 0.5 MPa$\sqrt{\text{m}}$ for the Al-Si-X alloys.

5.3 Evaluation of statistical characteristics of inhomogeneous particles

The present aluminum cast alloy AC4B-T6 contains three main types of inhomogeneous particles, namely, eutectic Si and Fe compounds and porosity. Surrounding an irregular cross section with a smooth convex curve as shown in **Figure 12**, the area is defined as $area_A$. The values of r are obtained from $area_A$ by the following equation [12, 37]:

$$r = \frac{\sqrt{area_A}}{\sqrt{\pi}}. \qquad (42)$$

Figure 13 shows the measured M_{A0} values of eutectic Si and Fe compounds and porosity. The relationship between M_{V0} and M_{A0} is as follows [12, 37]:

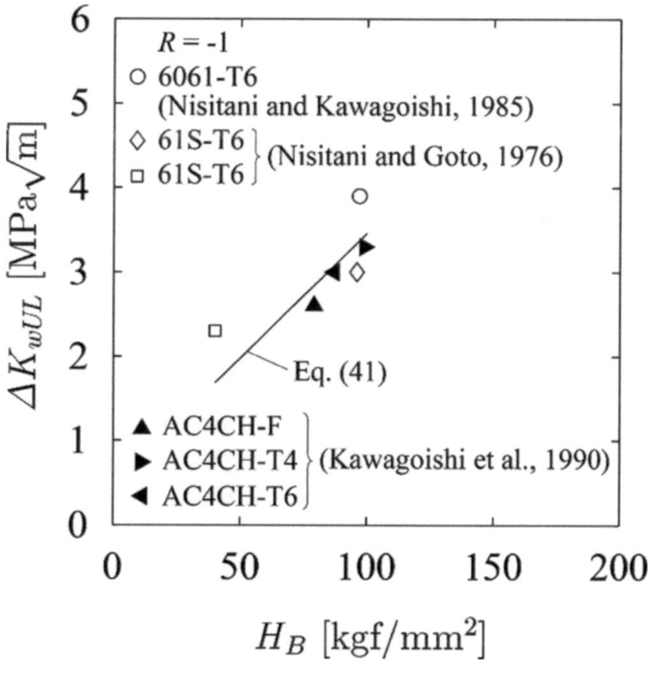

Figure 11.
Relation between ΔK_{wUL} and H_B.

$$M_{A0}(r_0) = 2\int_{\infty}^{r_0} \sqrt{R^2 - r^2}\, \frac{dM_{V0}}{dR} dR. \tag{43}$$

Here, $M_{A0}(r_0)$ is the number of cross-sectional particles in a unit area for which $r \geq r_0$ on a unit area. Considering the assumption that $M_{V0}(R_0)$ is given by Eqs. (1) and (2), the asymptotic characteristics of Eq. (42) are expressed by the following equation [12, 37]:

$$M_{A0}(r_0) \cong \sqrt{\frac{2\pi}{\nu}}\lambda \left(\frac{r_0}{\lambda}\right)^{1-\frac{\nu}{2}} \overline{N}_{V0} \exp\left\{-\left(\frac{r_0}{\lambda}\right)^{\nu}\right\}. \tag{44}$$

The line of Eq. (44) is drawn to best fit the M_{A0} values obtained by Eq. (43) to determine the values of $\overline{N}_{V0}$, ν, and λ.

Figure 14 shows the values of M_{V0} for porosity, Fe compounds, and eutectic Si. **Figure 15** shows the values of M_{S0} for the porosity, Fe compounds, and eutectic Si.

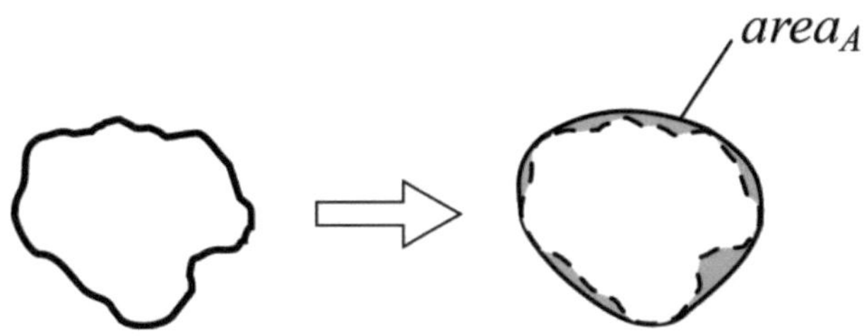

Figure 12.
Definition of $area_A$.

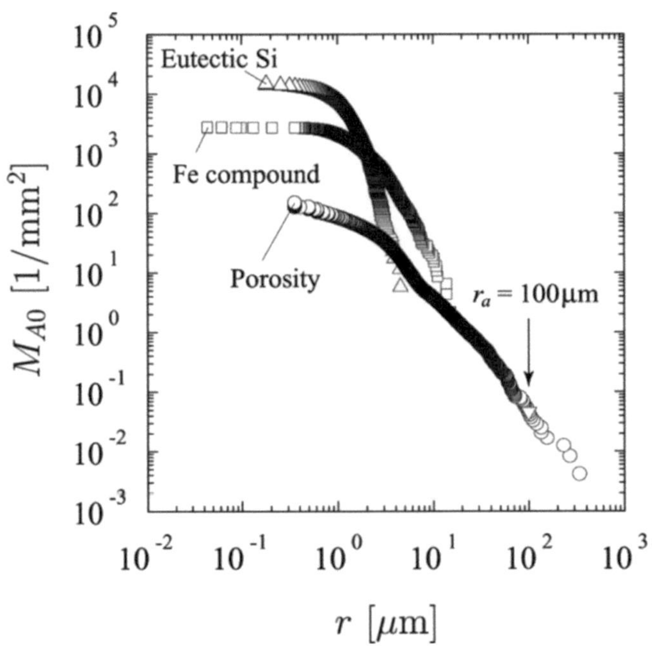

Figure 13.
M_{A0} of porosity and eutectic Si and Fe compounds.

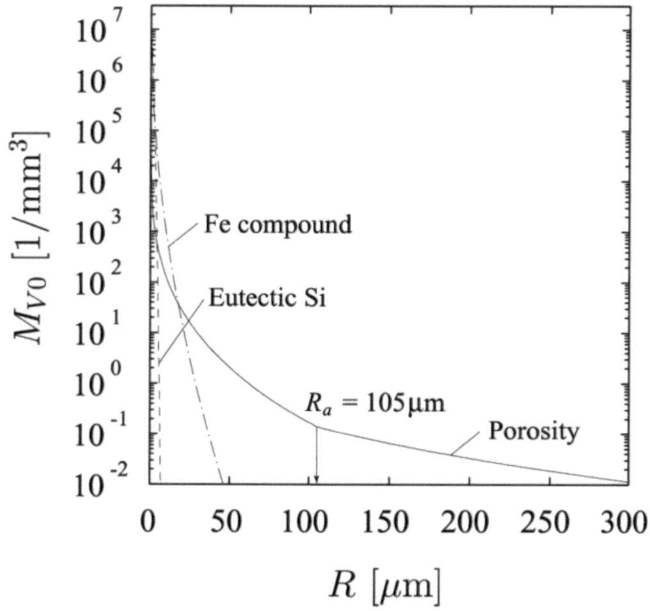

Figure 14.
Relation between M_{V0} and R.

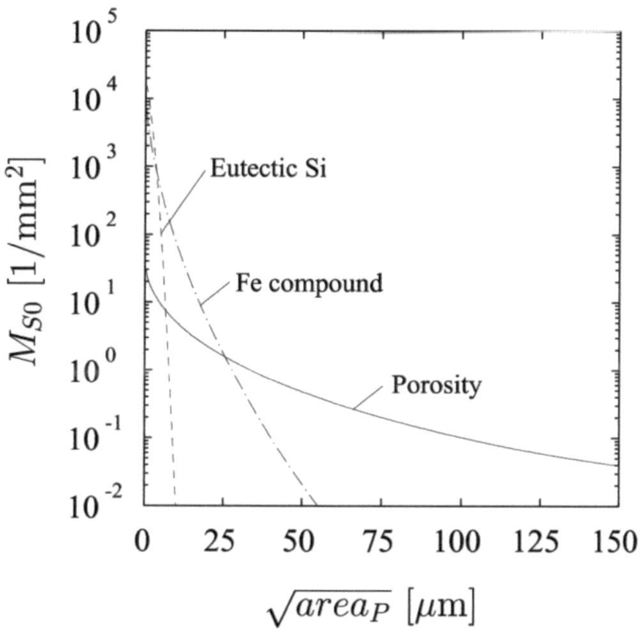

Figure 15.
Relation between M_{S0} and $\sqrt{area_P}$.

5.4 Evaluation of statistical characteristics of Vickers hardness of matrix without inhomogeneous particles

In this study, the Vickers hardness was measured at the position of 2.5–3.0 mm from the center on the circular cross section obtained by cutting the specimen grip under indentation load $P = 29.4$ mN in consideration with the stress distribution at the notch root. **Figure 16** shows the results plotted on a normal probability paper. The sample mean $\mu_{H_{VM}1}$ and sample variance $s^2_{H_{VM}1}$ were 91.8 kgf/mm^2 and 486.0 (kgf/mm^2)^2, respectively.

5.5 Evaluation of γ_m value

Because the values of R_m/p_m for the eutectic Si were much greater than those of the Fe compounds and porosity, as shown in **Table 3**, γ_m was calculated using only the eutectic Si. The eutectic Si was assumed to be a rigid body [7], and the following values were used for the calculation: E_M = 68 GPa, E_I = 105 GPa, and $\nu_M = \nu_I = 0.3$ [35]. Using R_m/p_m = 0.192, γ_m was determined to be 1.055.

5.6 Evaluation of $A_{npc\ S}$, $A_{npc\ R}$, and $A_{npc\ P}$ values

5.6.1 Evaluation of $A_{npc\ S}$ value

Because a microcrack often grows radially, it is approximated by the semielliptical crack shown in **Figure 17**.

$A_{npc\ S}$ in Eq. (20) can also be roughly evaluated. The approximation of the microcrack by the semielliptical macrocrack is such that the crack non-propagation

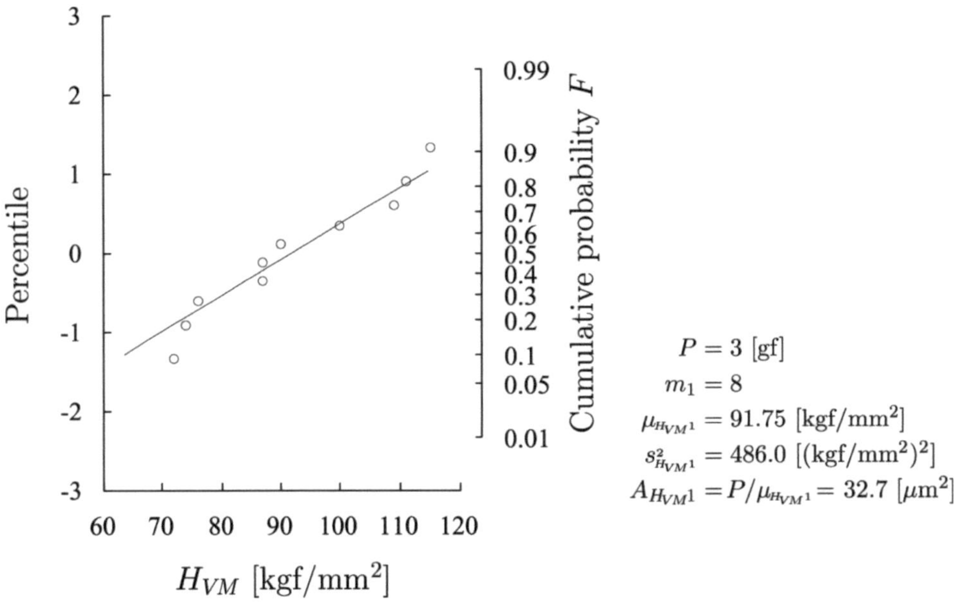

Figure 16.
Evaluation of Vickers hardness of matrix from normal probability paper.

Inhomogeneous particle		$\overline{N}_{V0}$ [1/mm^3]	ν	λ [μm]	R_m [μm]	p_m [μm]
Eutectic Si		8.73×10^6	1.6	1.04	0.932	4.86
Fe compound		2.20×10^7	0.5	0.10	0.200	3.57
Porosity	($R \geq 105\mu$m)	1.20×10^2	0.3	0.180	0.167	21.5
	($R < 105\mu$m)	1.00×10^5	0.3	0.0180		

Table 3.
Parameters of particle size distribution.

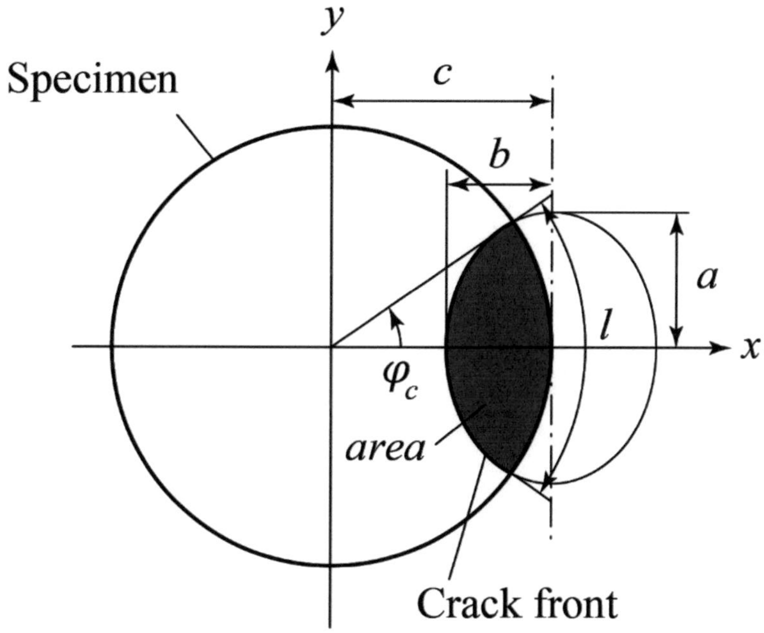

Figure 17.
Schematic illustration of the microcrack.

limits are equal. If the macrocrack is located near the notch root of radius ρ and it is sufficiently small, its non-propagation limit σ_{wdS} is given by the following equation:

$$\sigma_{wdS} = \frac{1.43(H_{VM} + 120)}{F_P \sqrt{area_P}^{1/6}}, \tag{45}$$

$$F_P = 0.968 - 1.021 \times 10^{-3} \frac{\sqrt{area_P}}{\rho}. \tag{46}$$

(σ_{wdS} is in MPa, H_{VM} is in kgf/mm^2, $\sqrt{area_P}$ is in μm, and ρ is in mm.)

Conversely, when the macrocrack is sufficiently large, ΔK_w is greater than ΔK_{wUL}. The macrocrack is thus treated as being large, and its non-propagation limit is categorized as σ_{w2}, which is given by the following equation:

$$\sigma_{w2} = \frac{13.0(H_{VM} + 16.7)}{F_P \sqrt{area_P}^{1/2}}. \tag{47}$$

(σ_{w2} is in MPa, H_{VM} is in kgf/mm^2, and $\sqrt{area_P}$: in μm.)

F_P is approximated to be 1. Because the average Vickers hardness of the matrix of the present AC4B-T6 is about 91.8 kgf/mm^2, the microcrack non-propagation limit σ_{w0} was estimated to be 160 MPa using Eq. (17). The non-propagating crack length of the present AC4B-T6 for ρ = 20 mm was about 60 μm when σ_n = 120 MPa. From these experimental results, the $l_{npc\ 0}$ of the present study was assumed to be 70 μm.

l_{npc} was set to achieve b/l = 0.4. Using c = 2.5 mm, $l_{npc\ 0}$ = 70 μm is equivalent to $\sqrt{area}_{npc\ 0}$ of 39.2 μm. **Figure 18** shows the relationship between l_{npc} and $1/\rho$.

5.6.2 Evaluation of $A_{npc\ R}$ and $A_{npc\ P}$ values

$A_{npc\ R}$ is a function of P_R and g_R; $A_{npc\ P}$ is a function of P_P and g_P. Because g_R and g_P can be calculated using Eqs. (26) and (33), respectively, only P_R and P_P need to

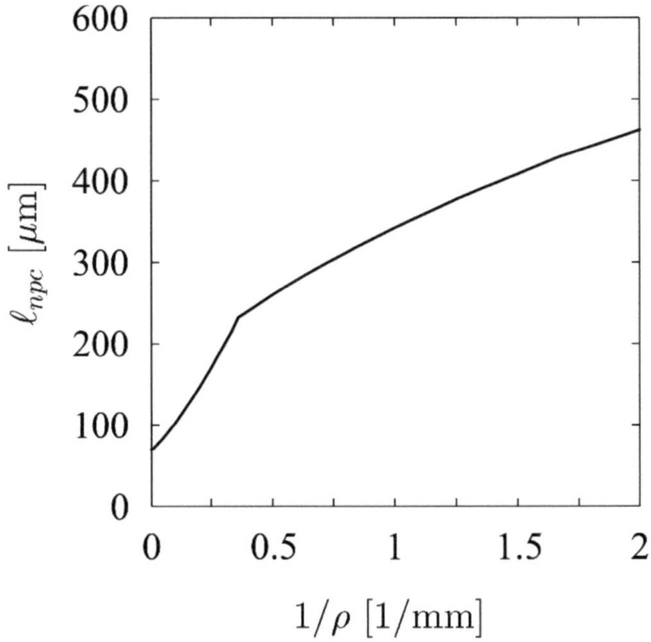

Figure 18.
Relation between l_{npc} and $1/\rho$.

be further examined. In this study, it is assumed that P_P = 0.3 kgf. Considering the difficulty in evaluating P_R, it is also assumed that $P_R = P_P$.

5.7 Comparison and examination of predicted and experimental results

The fatigue limit reliability of the notched specimen shown in **Figure 6** was predicted by the present method. The region in which the first principal stress is within the range of $\sigma_1^* = \sigma_1/\sigma_{\max}$ = [0.95, 1] at the center of the specimen was adopted as the control volume. In this case, the region was ring-like.

When H_B = 152 kgf/mm^2 is used, the ΔK_{wUL} = 5.06 MPa$\sqrt{\text{m}}$ is predicted from Eq. (41). Because the value of ξ for the present specimen is 0.167 (i.e., using d = 5 mm and t = 0.5 mm, as in Section 4.1), F is 0.754 [39]. In this case, the predicted value of σ_{w2} is 84.7 MPa. Considering that the experimentally determined value of σ_{w2} is 90 MPa, the prediction is confirmed to be good.

Figure 19 shows the fatigue limit reliability F_{σ_w}. The thick solid line represents the case of ρ = 2 mm, whereas the thin solid line represents the case of ρ = 0.3 mm. The value of F_{σ_w} for ρ = 0.3 mm suddenly changes from 0 to 1 when σ_w = 84.7 MPa, which is due to σ_{w1} and σ_{wd} being cut off by σ_{w2}. In other words, the inhomogeneous particles have almost no effect on the fatigue limit reliability in terms of initiating a fatigue crack. Instead, the eutectic Si actually strengthens the matrix.

Figure 20 shows the relationship between σ_w and $1/\rho$. The solid line represents 50% reliability, the broken line represents 90% reliability, the single-dotted chain line represents 99% reliability, and the open marks represent the experimental results. Because the fatigue limit obtained by the ordinary fatigue test is equivalent to 50% fatigue limit reliability, the solid line agrees well with the open marks. The little differences between the open marks of ρ = 0.3 and 0.1 mm and the solid line can be attributed to the fact that ΔK_{wUL} of the present AC4B-T6 was unknown and

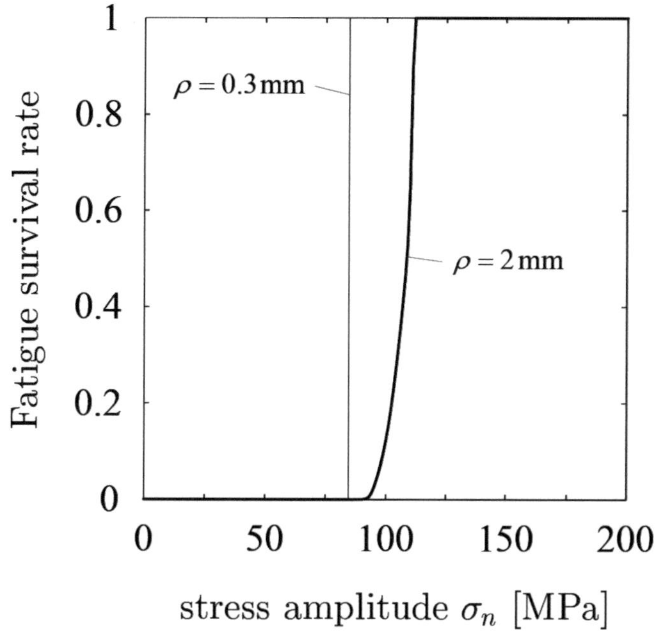

Figure 19.
Fatigue limit reliability F_{σ_w}.

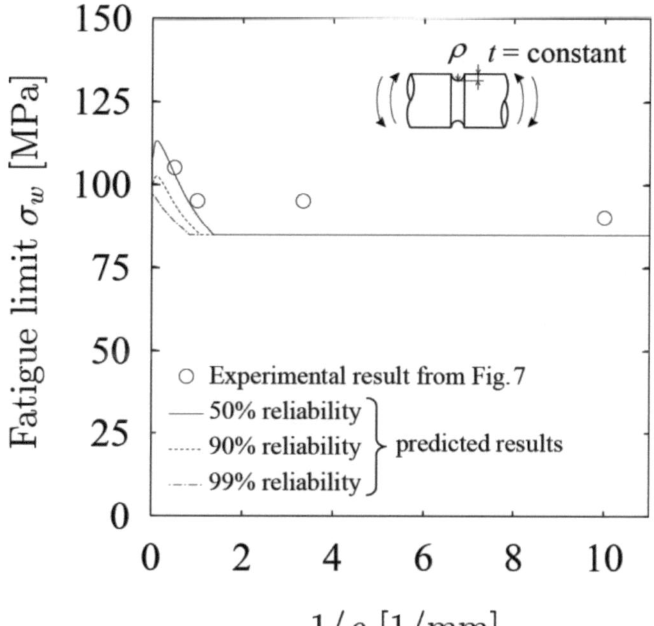

Figure 20.
Relation between σ_w and $1/\rho$.

the corresponding value for the of Al-Si-X alloy was used for predicting σ_{w2}. It is expected that an even better prediction accuracy would be achieved by using the true ΔK_{wUL}. Nevertheless, σ_w was well predicted, which validated the proposed method for notched AC4B-T6 specimens.

6. Conclusions

This study proposed a nondestructive method for predicting the fatigue limit reliability of notched specimens of a metal containing inhomogeneous particles densely. The method was applied to aluminum cast alloy JIS-AC4B-T6. Rotating-bending fatigue tests were performed on the notched specimens of AC4B-T6 with notch root radius ρ = 2, 1, 0.3, and 0.1 in order to examine the validity of the present method. Since the non-propagating macrocracks were observed along the notch root, the long macrocrack non-propagating limit σ_{w2} appears as the fatigue limit when ρ = 0.1 and 0.3 mm. On the other hand, since the non-propagating macrocrack was not observed when $\rho = 1$ and 2 mm, it can be said that the microcrack non-propagating limit σ_{w1} or the small macrocrack non-propagating limit σ_{wd} appears as the fatigue limit. The fatigue limits predicted by the present method were in good agreement with the experimental ones.

The method is not only convenient for use in predicting fatigue strength reliability for the reliable design of machine and structures, but it is also time effective and can be applied to the economic development of materials.

Author details

Tatsujiro Miyazaki[1], Shigeru Hamada[2] and Hiroshi Noguchi[2*]

1 Energy and Environment Program, School of Engineering, University of the Ryukyus, Okinawa, Japan

2 Department of Mechanical Engineering, Kyushu University, Fukuoka, Japan

*Address all correspondence to: nogu@mech.kyushu-u.ac.jp

References

[1] Hébert G, Dubé D, Tremblay R. Tensile and fatigue behaviour of thin-walled cast A383.0 components. Materials Science and Engineering A. 2012;**552**:89-96

[2] Fan KL, He GQ, Liu XS, Liu B, She M, Yuan YL, et al. Tensile and fatigue properties of gravity casting aluminum alloys for engine cylinder heads. Materials Science and Engineering A. 2013;**586**(1):78-85

[3] Hirsch J. Recent development in aluminium for automotive applications. Transactions of Nonferrous Metals Society of China. 2014;**24**(7):1995-2002

[4] Kelly A, Nicholson RB. Precipitation hardening. Progress in Materials Science. 1963;**10**:151-391

[5] Sjo Lander E, Seifeddine S. The heat treatment of Al-Si-Cu-Mg casting alloys. Journal of Materials Processing Technology. 2010;**210**:1249-1259

[6] Shaha SK, Czerwinski F, Kasprzak W, Friedman J, Chen DL. Effect of solidification rate and loading mode on deformation behavior of cast Al-Si-Cu-Mg alloy with additions of transition metals. Materials Science and Engineering A. 2015;**636**:361-372

[7] Kobayashi H, Ikeda H, Murakami Y. Super long life fatigue properties of Al-Si eutectic alloy by rotating-bending and tension-compression fatigue tests. Transactions of the Japan Society of Mechanical Engineers, Series A. 1996; **62**(594):347-355. (in Japanese)

[8] Wang QG, Apelian D, Lados DA. Fatigue behavior of A356/357 aluminum cast alloys. Part II—Effect of microstructural constituents. Journal of Light Metals. 2001;**1**(1):85-97

[9] Lea V-D, Morela F, Belletta D, Saintierb N, Osmondc P. Multiaxial high cycle fatigue damage mechanisms associated with the different microstructural heterogeneities of cast aluminium alloys. Materials Science and Engineering A. 2016;**649**(1): 426-440

[10] Laz PJ, Hillberry BM. Fatigue life prediction from inclusion initiated cracks. International Journal of Fatigue. 1998;**20**(4):263-270

[11] Debartolo EA, Hillberry BM. Effects of constituent particle clusters on fatigue behavior of 2024-T3 aluminum alloy. International Journal of Fatigue. 1998;**20**(10):727-735

[12] Miyazaki T, Kang H, Noguchi H, Ogi K. Prediction of high-cycle fatigue life reliability of aluminum cast alloy from statistical characteristics of defects at meso-scale. International Journal of Mechanical Sciences. 2008;**50**(2): 152-162

[13] Wang QG, Apelian D, Lados DA. Fatigue behavior of A356-T6 aluminum cast allys—Part I. Effect of casting defects. Journal of Light Metals. 2001; **1**(1):73-84

[14] Yia JZ, Leea PD, Lindleya TC, Fukuib T. Statistical modeling of microstructure and defect population effects on the fatigue performance of cast A356-T6 automotive components. Materials Science and Engineering A. 2006;**432**(1):59-68

[15] Gao YX, Yi JZ, Lee PD, Lindley TC. The effect of porosity on the fatigue life of cast aluminum-silicon alloys. Fatigue & Fracture of Engineering Materials & Structures. 2004;**27**:559-570

[16] JSME. JSME S 002-1994, Standard Method of Statistical Fatigue Testing. Tokyo, Japan: Japanese Society of

Mechanical Engineering; 1994. (in Japanese)

[17] ISO. ISO12107: Metallic Materials-Fatigue Testing-Statistical Planning and Analysis of Data. Geneva, Switzerland: International Organization for Standardization; 2012

[18] Murakami Y, Endo M. Effects of defects, inclusions and inhomogeneities on fatigue strength. International Journal of Fatigue. 1994;**16**:163-182

[19] Murakami Y, Tazunoki Y, Endo T. Existence of the coaxing effect and effects of small artificial holes on fatigue strength of an aluminum alloy and 70-30 brass. Metallurgical and Materials Transactions A. 1984;**15**(11):2029-2038

[20] Murakami Y. Metal Fatigue: Effects of Small Defects and Nonmetallic Inclusions. Oxford: Elsevier Science Ltd.; 2002

[21] Kobayashi M, Mutsui T. Prediction of fatigue strength of aluminum casting alloys by the $\sqrt{area}$ parameter model. Transactions of the Japan Society of Mechanical Engineers, Series A. 1996; **62**(594):341-346. (in Japanese)

[22] Ceschini L, Morri A, Morri A. Estimation of local fatigue behaviour in A356-T6 gravity die cast engine head based on solidification defects content. International Journal of Cast Metals Research. 2014;**27**(1):56-64

[23] Tajiria A, Nozakib T, Uematsub Y, Kakiuchib T, Nakajimac M, Nakamurac Y, et al. Fatigue limit prediction of large scale cast aluminum alloy A356. Procedia Materials Science. 2014;**3**: 924-929

[24] Roya MJ, Nadotb Y, Nadot-Martinb C, Bardinb P-G, Maijera DM. Multiaxial Kitagawa analysis of A356-T6. International Journal of Fatigue. 2011; **33**(6):823-832

[25] Miyazaki T, Noguchi H, Kage M, Imai R. Estimation for fatigue limit reliability of a metal with inhomogeneities under stress ratio $R = -1$. International Journal of Mechanical Sciences. 2005;**47**(2): 230-250

[26] Isibasi T. Prevention of Fatigue and Fracture of Metals. Tokyo: Yokendo Ltd.; 1967. (in Japanese)

[27] Nisitani H. Effect of size on the fatigue limit and the branch point in rotary bending tests of carbon steel specimen. Bulletin of the JSME. 1968; **11**(47):725-738

[28] Nisitani H, Goto T. Fatigue notch sensitivity of an aluminum alloy. Transaction of the Japan Society of Mechanical Engineers. 1976;**42**(361): 2666-2672. (in Japanese)

[29] Nisitani H, Kawagoishi N. Comparison of notch sensitivities in three age-hardened aluminum alloys. Transaction of the Japan Society of Mechanical Engineers, Series A. 1985; **51**(462):530-533. (in Japanese)

[30] Takao K, Nisitani H, Sakaguchi H. Relation between crack initiation process and notch sensitivity in rotating bending fatigue. Journal of the Society of Materials Science, Japan. 1980; **29**(325):982-987. (in Japanese)

[31] Kawagoishi N, Nisitani H, Tsuno T. Notch sensitivity of squeeze cast aluminum alloy in rotating bending fatigue. Transaction of the Japan Society of Mechanical Engineers, Series A. 1990; **56**(521):10-14. (in Japanese)

[32] Dowling NE. Notched member fatigue life predictions combining crack initiation and propagation. Fatigue & Fracture of Engineering Materials & Structures. 1979;**2**(2):129-138

[33] Miyazaki T, Noguchi H, Ogi K. Quantitative evaluation of the fatigue

limit of a metal with an arbitrary crack under a stress controlled condition (stress ratio $R = -1$). International Journal of Fracture. 2004;**129**:21-38

[34] Miyazaki T, Noguchi H, Kage M. Fatigue limit of steel with an arbitrary crack under a stress controlled constant with a positive mean stress. International Journal of Fracture. 2005; **134**:109-126

[35] Miyazaki T, Noguchi H, Kage M. Prediction of fatigue limit reliability of high strength steel with deep notch under mean stress $\sigma_m = 0$. International Journal of Fracture. 2011;**168**(1):73-91

[36] Miyazaki T, Noguchi H, Ogi K, Aono Y. Examination of fatigue characteristics of Aluminum cast alloy from meso-level consideration (2nd report, prediction for the fatigue limit reliability of plain specimen of metal containing different sorts of inhomogeneities under $R = -1$). Transactions of the Japan Society of Mechanical Engineers, Series A. 2005; **71**(712):1699-1707. (in Japanese)

[37] Hashimoto A, Miyazaki T, Noguchi H, Ogi K. Estimation of particle size distribution in materials in the case of spheroidal particles using quantitative microscopy. Journal of Testing and Evaluation. 2000;**28**(5):367-377

[38] Nisitani H. Method of approximate calculation of interference of notch effects and its application. Bulletin of the JSME. 1968;**11**(47):725-738

[39] Benthem JP, Koiter WT. Asymptotic approximations to crack problems. In: Method of Analysis and Solutions of Crack Problems. Leyden: Noordhoff International; 1973. pp. 131-178

Chapter 4

The Fracture Behavior of Pure and Hybrid Intraply Knitted Fabric-Reinforced Polymer Composites

Huseyin Ersen Balcioglu and Hayri Baytan Ozmen

Abstract

Due to the high synergistic effects of the components, hybrid composite materials are more advantageous than nonhybrid composite materials for advanced engineering applications. Additionally, knitted fabrics may have a different behavior than woven ones. Although the nonhybrid composites have only one reinforcing fiber type, the hybrid composites have multiple reinforcing fibers. In this chapter, fracture characterizations of laminated composites reinforced with intraply pure and hybrid knitted fabrics are experimentally and numerically investigated under different loading conditions. For this purpose, pure (100%) and hybrid fabrics (50–50%), which have 1×1 rib-knitted structure, were knitted by using glass and carbon fibers. Also, hybrid fabrics were knitted in three different widths in order to investigate the effect of knitting pattern width on the fracture toughness. Fracture toughness and energy strain release rates of pure and hybrid Arcan test specimens were determined under mode I (0°), mixed-mode I/II (30°, 45°, and 60°), and mode II (90°) loading conditions. Also, the J-integral method was used to determine the fracture toughness. Experimental and numerical results were found to be consistent. When the results obtained from pure and hybrid fabrics are compared, it is seen that hybridization had positive effects on the fracture strength of composite material compared to pure glass/epoxy material. Additionally, as the width of the pattern decreased, the fracture strength of the hybrid composites increased. In this respect, the hybridization processing should be done in the narrowest pattern width for higher resistance to fracture.

Keywords: fracture toughness, strain energy release rate, Arcan fracture test, pure and hybrid laminated composite, knitted fabric, J-integral method

1. Introduction

High strength to low weight ratio is a sought-after feature in the materials used in the structural elements of today's world. With the technological advances in recent years, composite materials are used in many industries, where durability and lightness are at the forefront, especially from the aerospace to automotive sectors. Polymeric composites have been used in many engineering applications due to their high strength in proportion to their weight, high stability, rigidity, superior corrosion, and fatigue resistance [1–3]. Woven or knitted fabrics of durable synthetic fibers such as glass, carbon, or aramid are used to reinforce polymer matrix composite. Woven reinforcement exhibits good stability in the warp and weft directions and offers the highest cover or yarn packing density in relation to fabric thickness [4]. Knitting is another technique of fabric formation for reinforcing. The fabric is formed by the inter-looping of yarn. The inter-looping of yarn can be done in two ways, namely, warp and weft knitting [5]. Complex lattice structures can be produced by local deformation of the loop in knitted structures. The loop that oriented through-thickness direction improves the out-of-plane mechanical properties of the structure. In addition, thanks to the perfectible geometry of the loop, high impact resistance and damage tolerance can be achieved. Textile-reinforced composites consist of a textile form as the reinforcement phase and usually a polymer for the matrix phase. 2D or 3D woven fabrics, knitted fabrics, stitched fabrics, braids, nonwovens, and multiaxial fabrics can be used as textile materials. Each of these textile forms has its own fiber architecture and combination of properties such as strength, stiffness, flexibility, and toughness, which are reflected on the composite performance to a certain extent [6]. A number of researchers have studied the damage strength of knitted fabric-reinforced composite structures under loading of tensile, compressive, fracture, and impact [7–10].

Lower cost, lower density, comparable specific strength, and better deformation capacity are the advantages of glass fibers as compared to carbon fibers. However, types of glass fibers have worse mechanical properties than carbon fibers, which is a limitation to the applications, especially when these materials are exposed to more severe stresses. One way to overcome this problem is hybridization, i.e., the combination of glass and carbon fibers, using a matrix compatible with both fibers, to obtain a composite material with satisfactory properties and lower cost. Hybridization with different fiber types within laminated composites increases the design space and opens up possible new engineering applications with optimized mechanical and functional properties. In addition, in the classical composites, the hybridization of carbon and glass fibers may cause a positive hybrid effect, which relies on the increase of carbon fibers failure strain, when compared to the pure carbon-based composites [11]. Generally, as far as the strength of the carbon/glass hybrid is concerned, the higher the volume percentage of carbon fibers is, the stronger the hybrid laminate becomes. The reason is that the carbon fibers are very tough and stiff, while the glass fibers are less stiff and less durable. In carbon/glass fiber-reinforced composites, an advantageous hybrid effect is also observed, which consists in increasing the carbon fiber failure strain when compared to the pure carbon fiber. In the literature, there are studies investigating the effects of carbon/glass hybridization on mechanical behavior. Tabrizi et al. [12] have investigated damage evolution in carbon/glass fiber hybrid composites with various stacking sequences under pure bending and tensile loading conditions. Swolfs et al. [13] concluded that the effect for tensile failure strain is well established, with a typical range of 10–50% for traditional hybrid composites such as carbon/glass. Wisnon et al. [14] have investigated hybrid effects on thin ply carbon/glass unidirectional laminates. Test results showed that the magnitude of the hybrid effect depends on the ply thickness. Dong and Davies [15] have studied the mechanical properties of the hybrid composites reinforced with the glass and carbon fibers. Naito and Oguma [16] have investigated tensile properties and fracture behavior of carbon/glass hybrid thermoplastic composite rods consisting of unidirectional PAN-based carbon fiber, braids of E-glass glass fibers, and thermoplastic epoxy matrix.

Composite materials used for structural purposes can be damaged during manufacturing, assembly, and usage of them. These damages can cause breaking of the materials under environmental effects and external loadings. One of these damages is crack onset and fracture, which depends on crack formation. Fracture, which is precarious for composite structures, can cause loss of life and property. Thus, the fracture analysis of the composite materials, especially focusing on the growth of defects that occur during the service that leads to destruction, is vital to the safety of the composite structures. The value of fracture toughness of composite materials strongly depends on three loading states at the end of the crack tip as the tensile opening mode (mode I), the in-plane shear mode (mode II), and the out-of-plane shear mode (mode III). However, fracture state does not form in pure mode I or mode II in the fiber-reinforced composite materials due to combined loading or anisotropy of composite structures [17]. For this reason, the study of the mixed-mode interlaminar fracture toughness is very important. Zhao et al. [18] have studied interlaminar fracture toughness of hybrid woven carbon-Dyneema composites with different hybridization schemes. The results showed that hybridization improves both mode I and mode II fracture toughness of carbon-Dyneema interfaces. Bienias et al. [19] have investigated interlaminar fracture toughness of woven glass and carbon-reinforced multidirectional fiber metal laminates under mixed-mode (mode I/II) loading. Jung and Kim [20] have investigated the fracture toughness of carbon-glass/epoxy interply hybrid composite under mode I loading condition. Saidane et al. [21] have investigated mode-I interlaminar fracture toughness of flax; glass and hybrid flax-glass fiber are woven composites by using a double cantilever beam test method. Swolf et al. [22] have studied translaminar fracture toughness of woven carbon/glass hybrid composites under impact loading.

Hybridizing two or more reinforcement materials within a matrix seeks to enhance the advantages of the reinforcing constituents and lessen the effect of the less desirable characteristics. In this study, the fracture toughness of pure and intraply hybrid knitted fabric-reinforced laminated composite plates have been investigated, experimentally and numerically. For this purposes, hybrid fabrics were knitted in a 1 × 1 rib-knitted structure by using glass and carbon fibers with equal weight carbon/glass fibers (50–50%). In order to investigate the effect of the knitting pattern width on the fracture behavior, the reinforcing hybrid fabrics were knitted at three different widths, such as 50, 25, and 12.5 mm. Arcan test apparatus was used to define mode I (opening mode), mode II (shearing mode), and mode I/II (mixed-mode) fracture toughness of test specimens. Also, fracture toughness for all composite samples was numerically determined in finite element analysis by using the J-integral method.

2. Materials and methods

2.1 Knitting process of hybrid fabric

Knitting is primarily classified as weft knitting and warp knitting. This classification is based on the direction of movement of yarn with respect to the direction of fabric formation. If the yarns run in the width or crosswise direction with reference to the direction of fabric formation during knitting, then the process of knitting is called weft knitting. The yarns in the knitted structure are just like weft yarns in woven fabrics. The weft-knitted fabrics made with one set of needles arranged in the grooves on one needle bed are called single jersey fabrics or plain knitted fabrics. In the experimental study, intraply hybrid reinforcement fabrics, which have 1×1 rib knitting structure, were knitted in a V-bed semi-automatic knitting machine. For this aim, 2400tex E-glass fibers and 3K carbon fibers were used as a knitting reinforcement element.

A loop is called a face loop or back loop according to the direction of the passing of one loop through another one during inter-looping (**Figure 1a**). A course is a horizontal row of loops produced by all the adjacent needles during the same knitting cycle. A wale is a vertical column of loops made by the same needle in

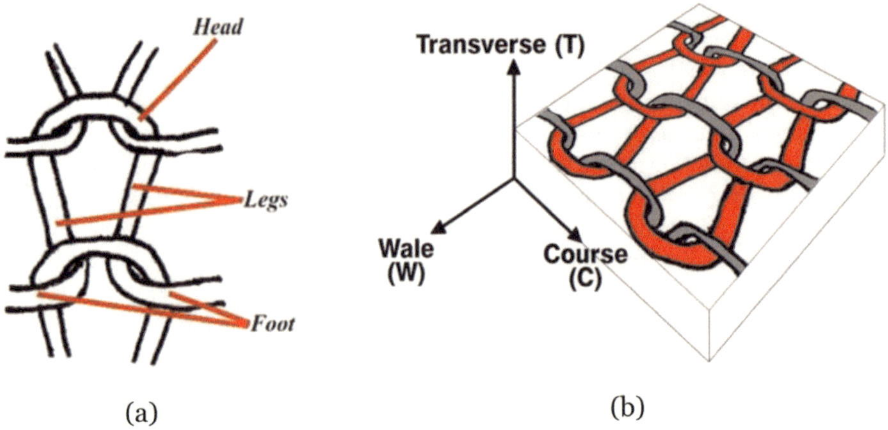

Figure 1.
(a) A knitted loop and (b) wale and course directions of loops.

Fiber type	Pattern width (mm)		
	12.5	25	50
Glass	4	8	15
Carbon	4	9	17

Table 1.
The row numbers of glass and carbon fibers for desired pattern widths.

successive knitting cycles. The direction of course and wale in weft-knitted fabric is shown in **Figure 1b**.

The row numbers of glass and carbon corresponding to the considered width are shown in **Table 1**. The average width of a single row is around 2.8–3.3 mm in the scope of this study. The average weight of hybrid and non-hybrid reinforcement fabrics was 730 g/m^2. The thickness of knitted fabric is approximately 2.7 mm.

2.2 Manufacturing of hybrid laminated composites

The matrix material was procured from Duratek Epoxy and Polyurethane Systems in Turkey. Hybrid laminated composite materials having four laminas were produced by hand lay-up methods. After all, laminas were saturated with epoxy resin; semi-product laminated composites were cured at 100°C under pressure of 8 MPa for 100 min, by using temperature-time-pressure-controlled hydraulic press.

After this process, the composite plates were cooled to room temperature under the same pressure to avoid warping effects. The fiber volume fractions for hybrid

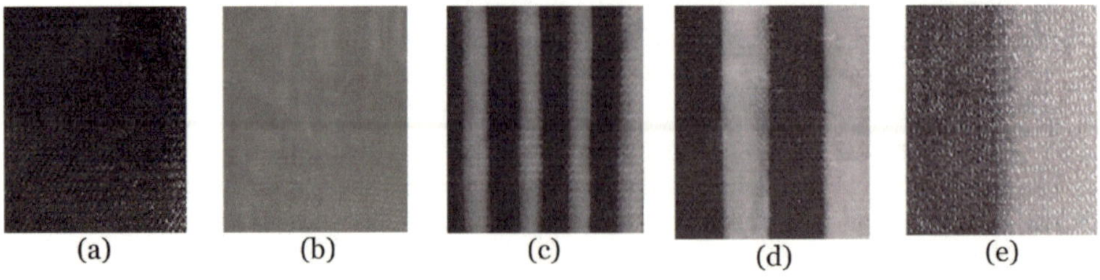

Figure 2.
Manufactured knitted fabric laminated composites (a) pure carbon/epoxy and (b) pure glass/epoxy fabric, and hybrid carbon-glass/epoxy composites with (c) 12.5, (d) 25, and (e) 50 mm pattern widths.

carbon-glass/epoxy laminated composites were determined as 55% approximately. In this study, the hybridization process is carried out using reinforcing fabrics knitted with different types of fibers on the same layer [23–25]. During stacking of layers, the same type of fibers was brought one on the top of another. The photographic representation of produced knitted hybrid composites is shown in **Figure** 2.

3. Experimental study

3.1 Determination of fracture toughness

The fracture toughness of pure and hybrid knitted laminated composites was determined for mode I (0°), mode I–II (30°, 45°, and 60°), and mode II (90°) by using modified Arcan test apparatus. In this context, Arcan test samples were cut with a CNC router machine by using 3 mm cutter blade (**Figure 3**). After cutting, crack having 4 mm was created on the Arcan test sample by using a jigsaw, which had 0.6 mm diameter. Hybrid composite specimens had two different reinforcement materials like glass and carbon fibers in the same layer. Therefore, crack onset in a different reinforcement material may occur in a different shape under load. Crack in the Arcan test sample having the same knitting pattern width was varied in two different forms to investigate crack onset mechanism in glass and carbon fibers.

In the first form, the crack was opened to glass fiber side and made to move toward the carbon fiber side. In the other form, the crack was opened to carbon fiber side and made to move toward the glass fiber side (**Figure 3c**, **d**). Linear elastic

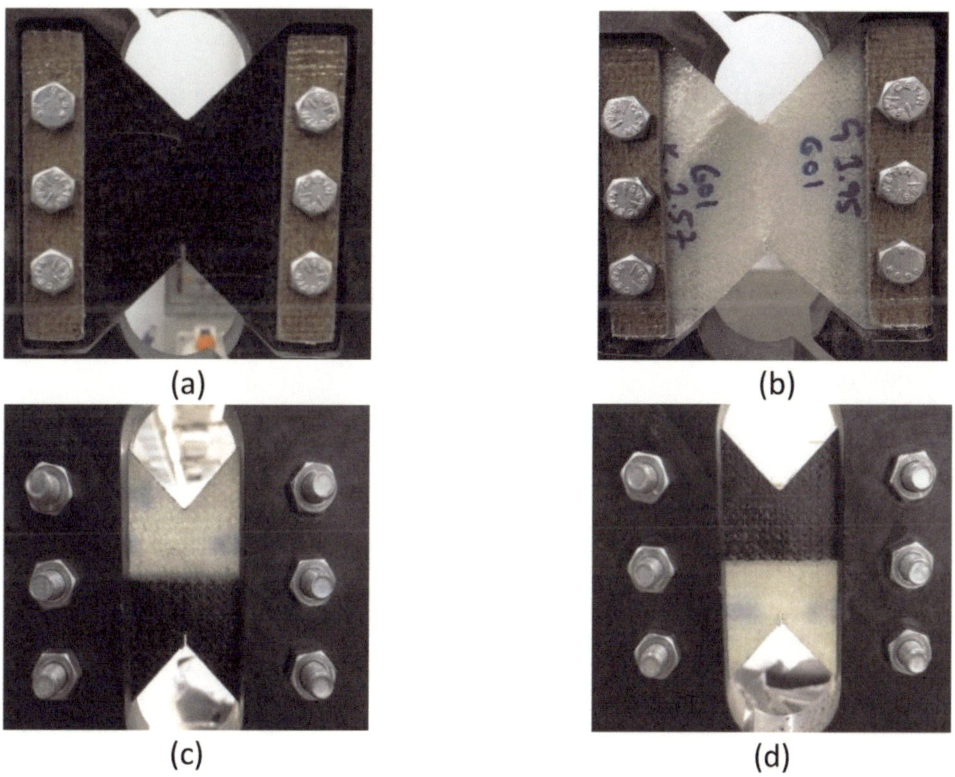

Figure 3.
Arcan test specimen (a) pure carbon/epoxy, (b) pure glass/epoxy, and hybrid carbon-glass/epoxy having (c) carbon side crack and (d) glass side crack.

fracture mechanics (LEFM) has been found as a useful tool for the investigation of cracks in composite materials. The purpose of fracture toughness testing is to determine the value of the critical stress intensity factor or plane strain fracture toughness K_C. The values of fracture toughness for the opening mode (mode I), tension/shearing mode (mixed-mode I/II), and shearing mode (mode II) were found using the following formulas. The stress intensity factor, K_C, at the tip of the crack for the Arcan test tension specimens is given by Eq. (1):

$$K_C = \frac{P_c\sqrt{\pi a}}{wt} f\left(\frac{a}{w}\right) \tag{1}$$

where P_c is the fracture load, a is the crack length, w is specimen width, t is the specimen thickness, and $f\left(\frac{a}{w}\right)$ is a geometrical factor. ASTM D5045 gives some guidance for plane strain fracture toughness and strain energy release rate [26]. ASTM D5045 is a reference in the literature for the studies, which are about the fracture behavior of fiber-reinforced polymer composites [27–29]. K_I (opening mode stress intensity factor) and K_{II} (shearing mode stress intensity factor) state the severity of the crack tip environment, and it is logical to characterize resistance to fracture by a critical value, that is, K_{IC} and K_{IIC} (Eq. 2). When the applied normal stress reaches the failure stress, the stress intensity factors K_I and K_{II} become the critical stress intensity factor (K_{IC} and K_{IIC}), which is taken as the fracture toughness of the composite specimens [30, 31]. The stress intensity factors ahead of the crack tip for Arcan test sample are calculated by using Eq. (2) [32–34]:

Mode I fracture toughness Mode II fracture toughness

$$K_{IC} = \frac{P_{max}\sqrt{\pi a}}{wt} f_I\left(\frac{a}{w}\right) \qquad K_{IIC} = \frac{P_{max}\sqrt{\pi a}}{wt} f_{II}\left(\frac{a}{w}\right) \tag{2}$$

The geometric factor formulas $f_I\left(\frac{a}{w}\right)$ and $f_{II}\left(\frac{a}{w}\right)$ used to calculate the K_I and K_{II} were provided by Eqs. (3–4) [32]:

$$f_I\left(\frac{a}{w}\right) = 182.12\left(\frac{a}{w}\right)^4 - 293.81\left(\frac{a}{w}\right)^3 + 187.87\left(\frac{a}{w}\right)^2 - 51.492\left(\frac{a}{w}\right) + 6.1137 \tag{3}$$

$$f_{II}\left(\frac{a}{w}\right) = -18.622\left(\frac{a}{w}\right)^4 + 36.753\left(\frac{a}{w}\right)^3 - 25.182\left(\frac{a}{w}\right)^2 + 7.759\left(\frac{a}{w}\right) + 0.0944 \tag{4}$$

For the mixed-mode loading effective fracture toughness, K_{eff} is calculated by Eq. (5) [35, 36]:

$$K_{eff} = \sqrt{K_I^2 + K_{II}^2} \tag{5}$$

where K_I and K_{II} are components of the fracture toughness in mode I and mode II directions. To calculate the values of mode I and mode II and total mixed-mode components of knitted fabric-reinforced composites, the material anisotropy should be taken into account.

3.2 Determination of strain energy release rate

The energy release rate (G) is defined as the amount of energy released per unit of the new fractured area formed due to cracking. The energy release rate is also defined as the crack extension force. A simple procedure using energy concepts is utilized to develop an analytical description of the crack extension force. The energy release rates for orthotropic material with the crack line parallel to the principal orthotropic direction which coincides with the fiber orientation can be calculated by Eq. (6):

Mode I strain energy release rate	Mode II strain energy release rate
$G_I = \frac{K_I^2}{E_I}$	$G_{II} = \frac{K_{II}^2}{E_{II}}$

(6)

where E_I and E_{II} are effective moduli for orthotropic materials. In order to apply the linear elastic fracture mechanics, the test sample must have some conditions. One of these is a load-displacement curve of the cracked test sample, which must show the linear elastic material behavior. Another requirement is that the strain state in the crack tip is known. When a material with a crack is loaded in tension, the materials develop plastic strains as the yield stress is exceeded in the region near the crack tip. Material within the crack tip stress field, situated close to a free surface, can deform laterally because there can be no stresses normal to the free surface. The state of stress tends to biaxial, and the material fractures in a characteristic ductile manner, with a 45° shear lip being formed at each free surface. This condition is called "plane stress," and it occurs in relatively thin bodies where the stress through the thickness cannot vary appreciably due to the thin section. Knitted fabric-reinforced composite materials conform to the conditions of thin plates. Therefore, the plane stress state occurs at the crack tip. E_I and E_{II} according to the plane stress state are expressed in Eq. (7) [32, 37]:

$$E_I = \frac{\sqrt{2E_wE_c}}{\sqrt{\sqrt{\frac{E_w}{E_c}} + \frac{2\vartheta G_{wc}+E_w}{2G_{wc}}}}, E_{II} = \frac{\sqrt{2}E_w}{\sqrt{\sqrt{\frac{E_w}{E_c}} + \frac{2\vartheta G_{wc}+E_w}{2G_{wc}}}} \tag{7}$$

where E_w is elasticity modulus in the wale direction and E_c is elasticity modulus in the course direction. In order to determine the mixed-mode strain energy release rate value, Eq. (8) is utilized [32]:

$$G_C = G_I + G_{II} \tag{8}$$

where similar to the effective fracture toughness formula, the values of G_I and G_{II} represent the energy release values in mode I and mode II directions.

4. Finite element analysis

In fracture analysis of composite materials, J-integral method expresses the stress energy release rate or work (energy) per unit fracture surface area. The J-integral defines the plastic stress and strain intensity in a manner similar to the fracture toughness (K) parameter, which represents the stress intensity of the surrounding elastic field, in the crack vicinity. The J-integral depends on stress, strain, crack size, and the geometry of the crack and body. The expression of J in the 2D form can be given by Eq. (9). It assumes that the crack lies in the global Cartesian x(u)–y(v) plane:

$$J = \int_{\Gamma}^{1} \left(W n_i^{\varepsilon} - n_i^{\varepsilon} \sigma_{ij} \frac{\partial u_j}{\partial x_1} \right) d\Gamma_{\varepsilon} \tag{9}$$

where Γ is a contour around the crack in the Cartesian coordinate system, W is stress field energy density, and x_i, u_i, and σ_{ij} are the spatial coordinates, the displacements, and the Cauchy stresses, respectively. n_i^ε are the components of the unit outward normal vector on Γ_ε. The coordinate system and typical J-integral paths were illustrated in **Figure 4**. For a linearly elastic solid, the strain energy density is expressed in Eq. 10:

$$W = W(x, \varepsilon) = \frac{1}{2}\sigma_{ij}\varepsilon_{ij} = \frac{1}{2}E_{ijkl}(x)\varepsilon_{ij}\varepsilon_{kl} \quad (10)$$

where $E_{ijkl}(x)$ are the Cartesian components of the fourth-order tensor expressing the generalized Hooke's law, ε is the strain tensor, and ε_{ij} are its Cartesian components [38]. Notice that the line integral shown in **Figure 4** is evaluated in a counterclockwise direction beginning on the lower crack surface and ending on the upper crack surface. The J-integral approach is a measure of the energy release rate associated with crack onset. One of the advantages of using J-integral is that under quasi-static conditions, it is equal to the energy release rate G for linear elastic materials [39]. Assuming linear elastic conditions prevailing along the integration path Γ, the numerical value of the J-integral can be related to the fracture toughness in plane stress condition as in Eq. (11) [40]:

$$J = \frac{K_C^2}{E} \quad (11)$$

The numerical analysis was performed in the commercial finite element software ANSYS Workbench by the use of a quasi-static rate-independent J-integral method, which can be used to determine the pure mode and mixed-mode fracture toughness. Eight-node quadrilateral plane elements with two degrees of freedom per node were used to model the cracked test specimens. The element has plasticity, creep, swelling, stress stiffening, large deflection, and large strain capabilities. The mesh was refined around the crack tip so that the smallest element size found in the crack tip elements was approximately 0.2 mm. The numerical model consists of 29,600 nodes approximately. An implicit solver was used for the finite element analysis. Implicit solutions are based on quantities calculated in the previous time step (backward Euler time scheme), which means even for large time steps the solution remains stable (unconditionally stable) [41]. In modeling fracture mechanics for laminated composite where both tensile and shear failure are common, a fracture criterion for predicting mode I, mode II, and mixed-mode I/II fracture onset is needed. Crack tip opening displacement test or CTOD is one of a

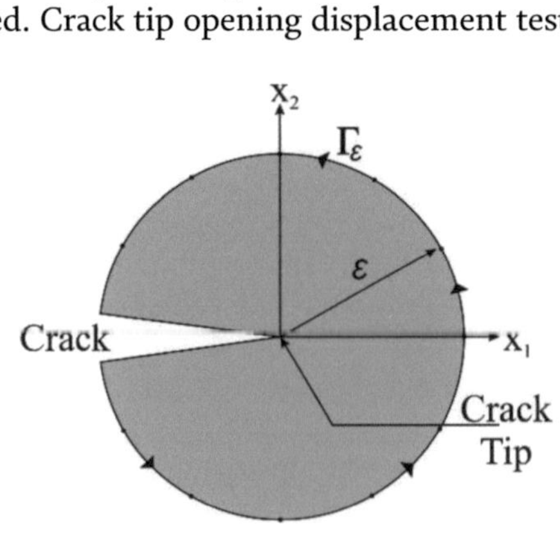

Figure 4.
The coordinates and typical paths to evaluate the J-integral.

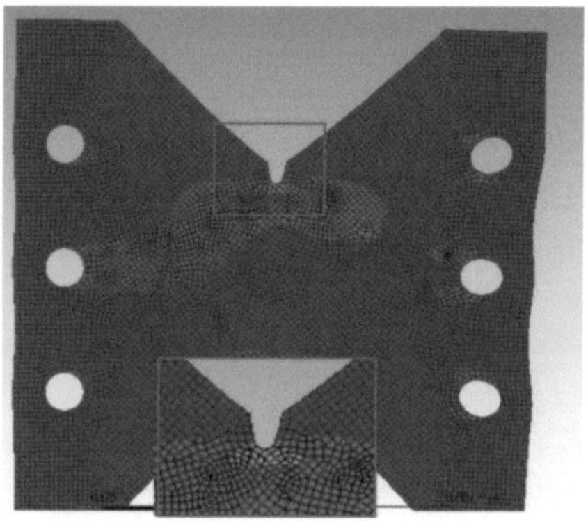

Figure 5.
The state of the finite element model after the analysis.

family of fracture criteria that measures the resistance of a material to growing a crack. In this context, CTOD failure criteria were employed for the crack onset in finite element analysis. There are two elastic-plastic parameters widely accepted by the fractured community; J-integral and CTOD [42, 43]. In case of LEFM, the elastic calculation for the CTOD can be expressed in Eq. (12):

$$\text{CTOD} = \frac{4}{\pi}\frac{K}{\text{E}\sigma_{ys}} \tag{12}$$

where K is the stress intensity factor, E is the effective modulus, and σ_{ys} is the uniaxial yield stress of the composite material.

The connection between the fixture and specimen is idealized by a rigid and continuous joint, based on the fact that the fixture and pins used in Arcan tests are relatively rigid compared to the specimen. Thus in the finite element analysis, the specimen-fixture system was treated as one continuous solid with two regions of different thickness and material properties. The fixed boundary condition was used, and distributed loads were assigned for the numerical model. **Figure 5** shows the state of the finite element model after the analysis.

5. Results and discussion

5.1 Results of Arcan fracture toughness test

The fracture tests were carried out using the Arcan test apparatus for 0°, 30°, 45°, 60°, and 90° loading angles. **Figure 6** has presented the load-displacement curve of pure glass/epoxy and pure carbon/epoxy at different loading angles. Load-displacement graphs of hybrid composites are not given because they behave similarly to others. When the loading angle changes from mode I to mode II plane, the maximum damage load (P_C) has increased. In addition, when the loading angle increased, specimens showed more deformation under load due to the increasing shear tendency of the test specimens. The fracture test was repeated five times for pure mode I (0°), pure mode II (90°), and all mixed-mode (30°, 45°, and 60°), and the obtained average P_C values are given in **Table 2**.

The average P_C values of critical fracture loads were used to determine the fracture toughness (K) and strain energy release rates (G) for all fracture modes. Calculated fracture toughness K_I, K_{II}, and K_{eff} according to the crack side was given in **Tables 3** and **4**, respectively.

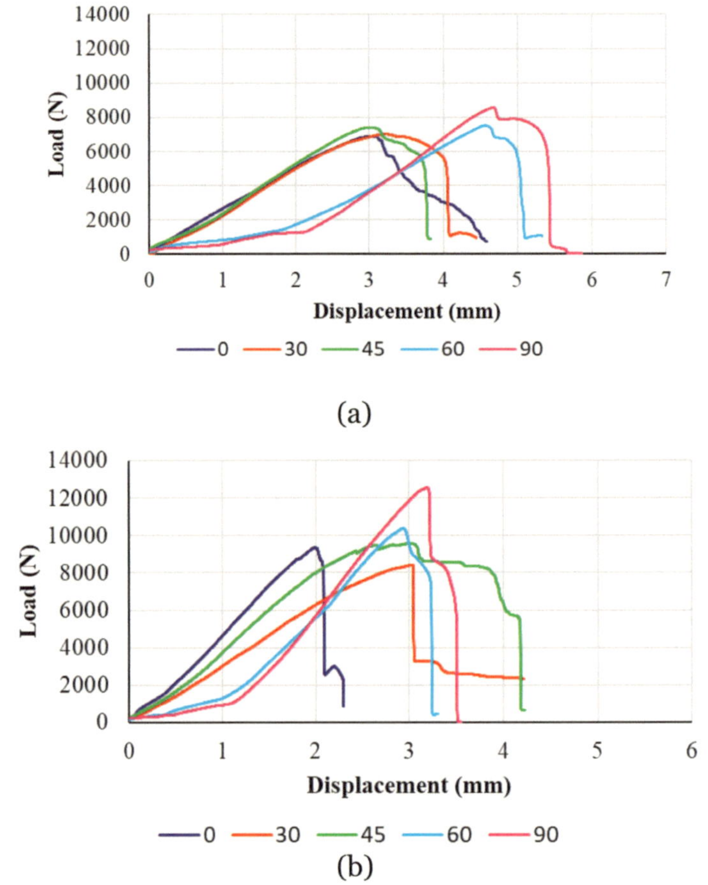

Figure 6.
Load-displacement curves according to loading angles (a) pure glass/epoxy and (b) pure carbon/epoxy.

Material type		**Loading angle**				
		0°	**30°**	**45°**	**60°**	**90°**
Glass side cracked	Pure glass/epoxy	6883	6902	6453	7576	8553
	Carbon-glass/epoxy (12.5 mm width)	8642	7970	8557	9345	9669
	Carbon-glass/epoxy (25 mm width)	8531	7791	8323	8990	8917
	Carbon-glass/epoxy (50 mm width)	7900	7373	7440	8405	8745
Carbon side cracked	Pure carbon/epoxy	9358	8406	9567	10366	12510
	Carbon-glass/epoxy (12.5 mm width)	8771	8772	9119	9816	9943
	Carbon-glass/epoxy (25 mm width)	8718	8187	9049	9453	9654
	Carbon-glass/epoxy (50 mm width)	8426	8178	8491	9356	9413

Table 2.
Average critical fracture loads P_C (N) for pure and hybrid knitted fabric laminated composites.

The calculated fracture toughness results showed that the loading angle, crack position, and pattern width directly affect the fracture behavior of the composite material. As loading angle increases from mode I to mode II, fracture toughness for each material type decreases. Applied load during mode I loading case forces the crack to open. So, damage occurs in the form of fiber and matrix fracture. In the

Material type		Loading angle				
		0°	30°	45°	60°	90°
Pure glass/epoxy	K_I	562.55	441.17	343.75	259.71	–
	K_{II}	–	63.46	83.40	109.18	161.30
	K_{eff}	562.55	445.71	353.72	281.73	161.30
Carbon-glass/epoxy (12.5 mm)	K_I	629.80	500.37	438.25	336.11	–
	K_{II}	–	70.28	110.97	146.10	158.59
	K_{eff}	629.80	505.28	452.08	366.50	158.59
Carbon-glass/epoxy (25 mm)	K_I	597.71	468.11	398.30	330.30	–
	K_{II}	–	67.03	103.33	135.53	146.70
	K_{eff}	597.71	472.89	411.49	357.02	146.70
Carbon-glass/epoxy (50 mm)	K_I	585.46	454.56	383.10	304.46	–
	K_{II}	–	64.14	94.61	130.11	141.90
	K_{eff}	585.46	459.06	394.61	331.10	141.90

Table 3.
Fracture toughness (MPa$\sqrt{mm}$) for carbon-glass/epoxy laminated composites with glass side cracked.

Material type		Loading angle				
		0°	30°	45°	60°	90°
Pure carbon/epoxy	K_I	664.64	532.40	494.83	360.78	–
	K_{II}	–	73.25	119.60	159.56	177.15
	K_{eff}	664.64	537.41	509.08	394.49	177.15
Carbon-glass/epoxy (12.5 mm)	K_I	635.60	523.76	467.58	352.70	–
	K_{II}	–	71.40	115.39	153.05	165.03
	K_{eff}	635.60	528.60	481.61	384.47	165.03
Carbon-glass/epoxy (25 mm)	K_I	619.57	505.11	436.68	345.04	–
	K_{II}	–	70.15	109.41	147.50	151.22
	K_{eff}	619.57	509.96	450.18	375.24	151.22
Carbon-glass/epoxy (50 mm)	K_I	599.85	479.02	415.84	337.96	–
	K_{II}	–	68.60	107.79	144.49	148.98
	K_{eff}	599.85	483.90	429.59	367.55	148.98

Table 4.
Fracture toughness (MPa$\sqrt{mm}$) for carbon-glass/epoxy laminated composites with carbon side cracked.

case of the opening mode, the damage occurs in the form of a matrix crack and subsequent fiber breakage. Also, the breakage occurs in a fast and brittle form, due to high-stress concentrations occurring at the crack end in the opening mode. During mode II loading, the applied load progresses the crack by shearing. As a result of shear deformation, the separation between laminas named delamination occurs. Damage of the brittle matrix material holding the lamina together allows the delamination to spread easily between the laminas. Thus, the material gains more ability to deform. The loads on the sample are transferred to the reinforcing hybrid fabric with increasing deformation, and the fiber structure is subjected to shear force. Due to the anisotropic behavior of the knitting structure, it can be seen from fracture test results that the shear strength of reinforcement fabric was higher than the tensile strength.

When fracture toughness values of pure glass and carbon fabric-reinforced composites were compared, fracture toughness of carbon/epoxy composites was found to be up to 43% higher than glass/epoxy. If a similar comparison is made for hybrid composites that had the same pattern width, the fracture toughness of the samples with carbon side cracked is 11% higher than for samples with glass side cracked. According to the results obtained from pure and hybrid composites, the crack on the carbon side has a tougher spreading mechanism than on the glass side.

Although all hybrid fabric-reinforced composites contain equal amounts of glass and carbon fiber, different fracture toughness values were obtained for the same loading angle and crack location. Glass and carbon knitting pattern widths of hybrid fabrics have affected the fracture toughness of the material. At the combination boundary of the glass and carbon fiber knitting, a new intermediate form is occurred by the interlocking of glass fiber and carbon fiber loops. This intermediate form increased the strength of the structure due to exhibited behavior that is as flexible as the glass fiber and as strong as the carbon fiber. Accordingly, the fracture toughness value of the material has increased by decreasing pattern width or in other words increasing the number of intermediate forms. When the fracture toughness values at the same loading angle of the samples having the carbon side crack were compared with regard to the pattern width, the samples having a pattern width of 12.5 mm have more toughness value up to 9 and 12%, respectively, than the samples with 25 and 50 mm pattern width. If the similar comparison was made for the samples having a crack on the glass side, it was seen that the samples with a pattern width of 12.5 mm have more toughness value up to 10 and 15% than those with a pattern width of 25 and 50 mm, respectively.

5.2 Results of finite element analysis

A numerical study was also performed by using ANSYS finite element program for all loading angles. Some mechanical test values, which required to create a finite element model of glass and carbon knitted fabric-reinforced composite structures, were determined experimentally, and the obtained results are given in **Table 5**. The elasticity modulus in the wale direction (E_w) and the course direction (E_c) and the tensile strength in the wale (T_w) and course direction (T_c) of laminated composites were determined according to ASTM D3039M standard [44]. Shear modulus (G_{wc}) was determined according to the ASTM D3518M-13 standard test method [45]. Compressive properties were determined according to the ASTM D3410-87 standard test method [46]. Wale and course direction compressive strength of composite specimens (C_w and C_c) were calculated by dividing the failure load to the cross-sectional area of the specimens in wale and course direction, respectively. The in-plane shear properties in the wale direction (S_{wc}) and in the course direction

Material type	E_w (MPa)	E_c (MPa)	G_{wc} (MPa)	T_w (MPa)	T_c (MPa)	C_w (MPa)	C_c (MPa)	S_{wc} (MPa)	S_{cw} (MPa)
Glass/epoxy	24105.84	19621.19	4160.16	127.82	117.80	98.61	81.94	40.33	36.21
Carbon/epoxy	40437.71	29891.17	5112.89	200.91	156.36	125.80	99.27	51.94	46.23

Table 5.
Mechanical properties of nonhybrid knitted fabric-reinforced composite specimens.

Material type		Loading angle				
		0°	30°	45°	60°	90°
Pure glass/epoxy	J_{int}	17.81	8.52	3.19	2.16	1.28
	G_c	14.98	9.38	5.88	3.68	1.06
Carbon-glass/epoxy (12.5 mm)	J_{int}	16.97	11.61	9.63	6.65	0.95
	G_c	18.78	12.05	9.59	6.22	1.02
Carbon-glass/epoxy (25 mm)	J_{int}	14.93	11.13	7.93	5.53	0.90
	G_c	16.91	10.56	7.94	5.91	0.88
Carbon-glass/epoxy (50 mm)	J_{int}	14.34	10.02	7.93	4.82	0.67
	G_c	16.23	9.95	7.31	5.08	0.82

Table 6.
Comparison of J-integral and strain energy release rate (G_c) of glass side cracked specimens.

(S_{cw}) of glass and carbon knitted fabric-reinforced composites were determined according to ASTM D 5379 standard by using V-notched test samples [47]. The S_w and S_c have been found by dividing of maximum load by the cross-sectional area of the samples. All the tests for the mechanical properties were done five times for each material structure in room temperature. The average results of these five tests were accepted as mechanical property values. When **Table 5** is investigated, it can be seen that the mechanical strength of carbon/epoxy composite is higher than the glass/epoxy composite. Such that, tensile properties of pure carbon/epoxy as E_w, E_c, T_w, and T_c are 67.75, 52.34, 57.18, and 32.73% higher than glass/epoxy, respectively. The compression strength of carbon/epoxy in wale and course direction are 27.57 and 21.15% higher than glass/epoxy, respectively. For the comparison of shear modulus and shear strength, it can be seen that G_{wc}, S_{wc}, and S_{cw} values of carbon/epoxy are 22.9, 28.78, and 27.66% higher than glass/epoxy, respectively.

From the physical point of view, the energy release rate is the most appropriate physical quantity to characterize the fracture behavior. For purely elastic materials, the energy release rate G is identical to the J-integral because there is no energy stored in the crack cavity. In linear elastic fracture mechanics, the J-integral coincides with total energy release rate, $J_{int} = G_c = G_I + G_{II} + G_{III}$, where G_I, G_{II}, and G_{III} are the energy release rates associated with the mode I, mode II, and mode III stress intensity factors. In this study, the energy release rate (G) is obtained by using experimental data in theoretical formulas.

The J-integral value is calculated by the ANSYS program with the aid of the finite element model. The comparisons of the J-integral and strain energy release rate values, which were obtained from experimental and numerical analyses, were given in **Tables 6** and **7** depending on the crack location. When the comparisons in **Tables 6** and **7** are examined, it is seen that experimental and numerical results are compatible with each other. However, for the samples having the same pattern width, the energy required to progress the carbon side crack is higher than the glass side crack at the same loading angle.

In linear elastic fracture mechanics, Eq. (11) is valid between the fracture stress intensity factor (K_C) and the J-integral value for plane stress and plane strain cases. During the analysis, if the thickness of the material is neglected, plane stress condition is applicable, and if it is included in the solution, the plane strain condition is applicable. Depending on the J-integral value obtained from finite element numerical analysis, fracture toughness was determined according to Eq. 13 for plane stress condition [30, 37, 48]:

$$K_J = \sqrt{J.E_{effective}} \tag{13}$$

A comparison of the experimentally obtained fracture toughness values ($(K_C)_{exp}$) and the numerical fracture toughness values obtained using J-integral ($(K_J)_{num}$) is given in **Tables 8** and **9** according to the crack location.

According to **Tables 8** and **9**, it can be said that the results are close to each other when numerical values are compared with experimental values. The maximum error value was 15% for the finite element analysis when the experimental values are taken as reference. This maximum error value indicates that the numerical model created for finite element analysis successfully converges to the fracture test condition.

Damage modes and stress distributions of laminated composites were given in **Figure 7** after experimental fracture damage and FEM analysis. Due to important stress concentrations around the notches in uniaxial tension, specimen fracture

Material type		Loading angle				
		0°	30°	45°	60°	90°
Pure carbon/epoxy	J_{int}	18.38	11.98	9.81	6.65	1.47
	G_c	20.91	13.64	12.17	7.20	1.28
Carbon-glass/epoxy (12.5 mm)	J_{int}	17.45	12.96	9.82	7.09	1.09
	G_c	19.12	13.19	10.89	6.84	1.11
Carbon-glass/epoxy (25 mm)	J_{int}	15.20	11.05	8.84	5.84	0.95
	G_c	18.17	12.28	9.51	6.52	0.93
Carbon-glass/epoxy (50 mm)	J_{int}	15.62	10.26	7.82	5.52	1.05
	G_c	17.03	11.05	8.66	6.26	0.90

Table 7.
Comparison of J-integral and strain energy release rate (G_c) of carbon side cracked specimens.

Material type		Loading angle				
		0°	30°	45°	60°	90°
Pure glass/epoxy	$(K_C)_{exp}$	562.55	445.71	353.72	281.73	161.30
	$(K_J)_{num}$	516.85	417.70	317.40	261.29	145.66
	% error	8.12	6.28	10.27	7.25	9.70
Carbon-glass/epoxy (12.5 mm)	$(K_C)_{exp}$	629.80	505.28	452.08	366.50	158.59
	$(K_J)_{num}$	639.68	551.30	515.75	400.53	172.45
	% error	1.57	9.11	14.08	9.29	8.74
Carbon-glass/epoxy (25 mm)	$(K_C)_{exp}$	597.71	472.89	411.49	357.02	146.70
	$(K_J)_{num}$	549.63	536.17	462.49	371.52	159.11
	% error	8.04	13.38	12.39	4.06	8.46
Carbon-glass/epoxy (50 mm)	$(K_C)_{exp}$	585.46	459.06	394.61	331.10	141.90
	$(K_J)_{num}$	499.28	517.96	437.08	377.01	127.37
	% error	14.72	12.83	10.76	13.87	10.24

Table 8.
Experimental and numerical fracture toughness values of glass side cracked specimens.

Material type		Loading angle				
		0°	30°	45°	60°	90°
Pure glass/epoxy	$(K_C)_{exp}$	664.64	537.41	509.08	394.49	177.15
	$(K_J)_{num}$	623.11	564.88	468.68	408.48	189.78
	% error	6.25	5.11	7.94	3.55	7.13
Carbon-glass/epoxy (12.5 mm)	$(K_C)_{exp}$	635.60	528.60	481.61	384.47	165.03
	$(K_J)_{num}$	607.15	597.83	541.68	426.50	180.12
	% error	4.48	13.10	12.47	10.93	9.14
Carbon-glass/epoxy (25 mm)	$(K_C)_{exp}$	619.57	509.96	450.18	375.24	151.22
	$(K_J)_{num}$	674.09	574.84	513.97	417.84	163.68
	% error	8.80	12.72	14.17	11.35	8.24
Carbon-glass/epoxy (50 mm)	$(K_C)_{exp}$	599.85	483.90	429.59	367.55	148.98
	$(K_J)_{num}$	524.25	553.78	484.21	406.19	160.65
	% error	12.60	14.44	12.72	10.51	7.83

Table 9.
Experimental and numerical fracture toughness values of carbon side cracked specimens.

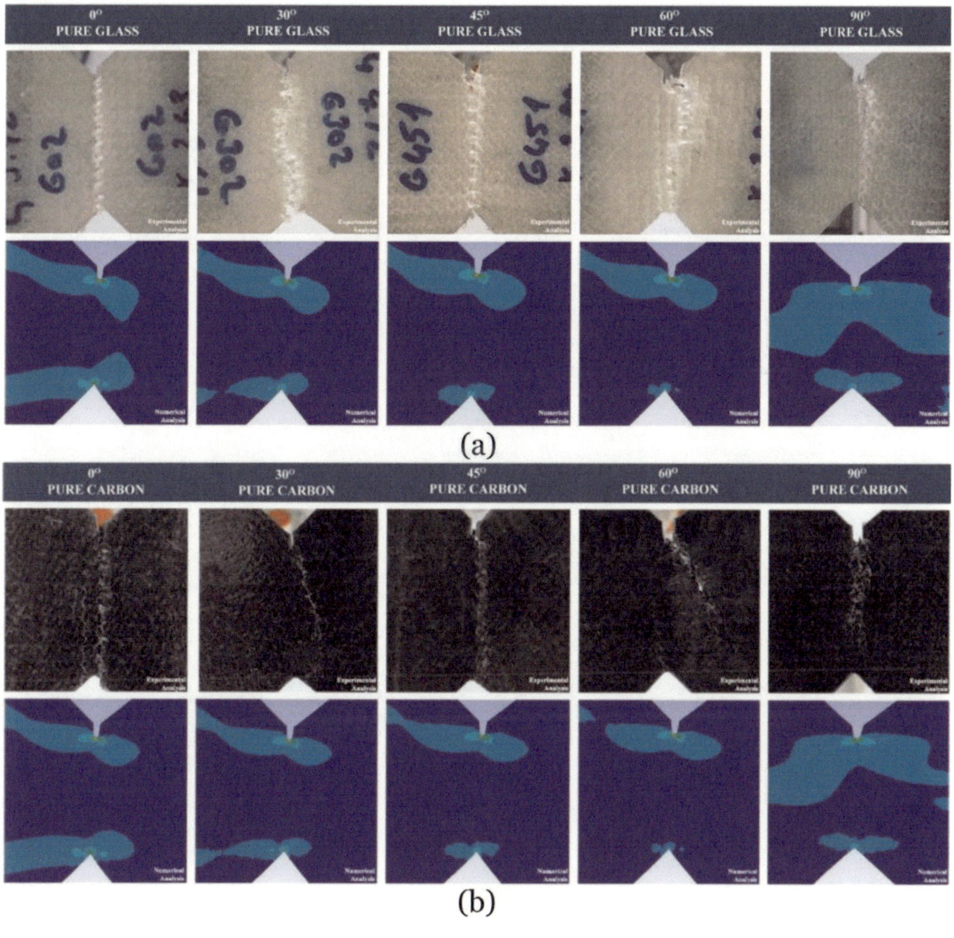

Figure 7.
Damage and stress distributions after experimental fracture and FEM analysis of laminated composites for (a) glass/epoxy and (b) carbon/epoxy.

generally occurred in the significant crack tip. As noticed from previous tests, the fracture mechanism consisted of one unique cracked interface identical for all glass/epoxy composite samples tested whatever the loading direction. In such a case, the crack propagated between the two loops of rows, which is in the direction of the wale, and the crack could not pass through the other loops. Although different crack onset mechanisms did not appear depending on the loading angle in the glass/epoxy specimens, FEM analyses have shown that the stress distributions in the crack region vary depending on the loading angle. Fractured carbon/epoxy composite samples presented a nearly horizontal cracked zone that was different from a plane surface for loading angle of 30° and 60°. During the experimental Arcan test, it was observed that the crack progress in the main delamination plane without any side cracking and branching. Fiber bending and breaking behind the crack tip were observed macroscopically in crack onset during the test.

The experimental and numerical analysis visual results of hybrid composites with 12.5 mm pattern width, which have the maximum fracture toughness values, are given in **Figure 8**. The crack propagates by the glass in **Figure 8(a)** and by carbon in **Figure 8(b)**. Von Mises stress distribution for different loading angles, which obtained from finite element analysis, is shown. The crack started from the

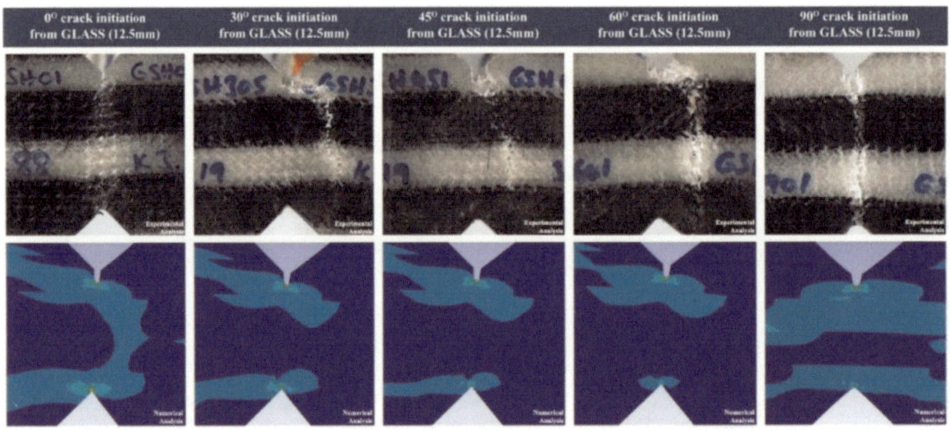

(a)

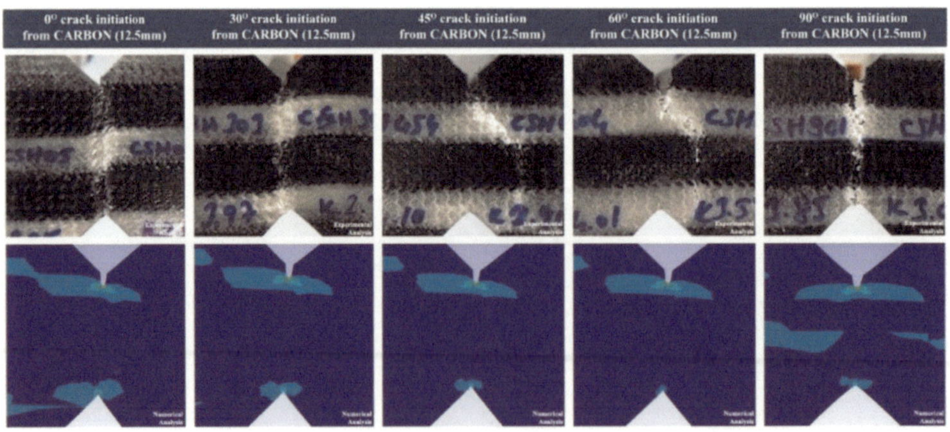

(b)

Figure 8.
The experimental and numerical visual results of hybrid composites with 12.5 mm pattern width having (a) glass side crack and (b) carbon side crack.

crack tip due to high-stress concentration at the crack tip, and it propagates to the other side by breaking fibers and/or fiber pull out. It can be clearly said that the numerical damage forms were obtained in the similar views of the experimental damage forms as illustrated in **Figure 8**. According to the results of numerical damage, Von Mises stresses show a vertical progression in the case of mode I, while a more horizontal progression occurs in the case of mode II.

6. Conclusion

This paper has presented the fracture behavior of pure and hybrid knitted fabric-reinforced laminated composites based on experimental and numerical analyses. In this context, the effect of crack location, loading angle, and pattern width on fracture behavior are examined. A modified version of the Arcan test fixture was employed to conduct a mode I, mode II, and mode I/II test. The obtained fracture test results of hybrid specimens are compared with the test results of pure glass/epoxy and pure carbon/epoxy samples. In addition, finite element models of cracked test specimens were created according to the data obtained from the mechanical tests. Fracture behaviors of hybrid composites were numerically analyzed using J-integral method. The concluding remarks in this study can be summarized as follow:

- According to the results obtained from the mechanical tests, knitted fabrics have been found to be an alternative to woven fabrics for reinforcing polymer composites. In addition, the test results show that the mechanical strength values change depending on the knitting direction and all mechanical test values are larger in the wale direction.

- The maximum and minimum fracture toughness value for carbon-glass/epoxy hybrid laminated composites was obtained in mode I and mode II loading conditions. The highly complex structure of the knitted fabric composites induces various toughening mechanisms. Fracture toughness behaviors of pure and hybrid composites varied in terms of loading angle. When, the loading angle increased from 0 (mode I) to 90 (mode II), the critical damage load increases. On the contrary, the fracture toughness and energy release rate decreases. The results indicated that the Arcan cracked specimen is tougher in tensile loading conditions and weaker in shear loading conditions.

- The rib-knitted fabric-reinforced composite shows different fracture toughness and energy release rate values for both crack progression directions despite the fact that damage images showed that the crack growth modes are different: in the wale direction, the crack followed the wavy surface of the fabric, and in the course direction, the majority of yarns is broken. In the wale direction, the major fracture mechanisms were the matrix deformations, leading to micro-cracks, which will branch in a network. In the course direction, the crack does not strictly follow the waviness of the fabrics but tends to grow through them. The main damage occurs by multiple fiber breakage. This phenomenon is supposed to be highly energy consuming because it implies events such as peel off, yarn bridging, and yarn failure.

- For all fracture tests of pure fabric-reinforced composite, carbon/epoxy specimens were much more resistant than glass/epoxy in terms of failure loads whatever the loading angle.

- As the width of the pattern increased, the fracture strength of the hybrid composites decreased. In this respect, the hybridization processing should be done in the narrowest pattern width for high resistance to fracture.

- In terms of crack locations, the progression of the crack in the glass-reinforced zone is more hazardous than the progress in the carbon reinforcing zone. During the assembly of carbon-glass hybrid composites, it is better to ensure that the bolt holes are opened on the carbon side if the bolts are to be used.

- When the fracture toughness values that were obtained experimentally and numerically are compared, it is seen that the results are consistent. In addition, in terms of fracture energy, experimentally obtained strain energy release rate (G) and numerical fracture energy (J-integral) values are similar. In this respect, the usability and validity of the J-integral method have been proven to simulate numerical fracture analysis of knitted fabric-reinforced laminated composites.

Author details

Huseyin Ersen Balcioglu[1*] and Hayri Baytan Ozmen[2]

1 Department of Mechanical Engineering, Usak University, Usak, Turkey

2 Department of Civil Engineering, Usak University, Usak, Turkey

*Address all correspondence to: ersen.balcioglu@usak.edu.tr

References

[1] Balcıoğlu HE, Sakin R, Gün H. The design of multi-sample flexural fatigue device and fatigue behavior of glass/epoxy laminated composites. Research on Engineering Structures and Materials. 2018;**4**:279-296. DOI: 10.17515/resm2018.67me0917

[2] Atay HY. Multi-functional materials for military aircrafts; radar absorbing and flame retardant composites. Research on Engineering Structures and Materials. 2016;**3**:45-54. DOI: 10.17515/resm2016.38ma0204

[3] Costa DMS, Loja MAR. On the characterization of the free vibrations behavior of multiscale composite plates. Research on Engineering Structures and Materials. 2016;**3**:27-44. DOI: 10.17515/resm2016.43st0602

[4] Karakoç A. A fiber network model to understand the effects of fiber length and height on the deformation of fibrous materials. Research on Engineering Structures and Materials. 2016;**2**:51-57. DOI: 10.17515/resm2015.17ma0825

[5] Ashraf W, Nawab Y, Umair M, Shaker K, Karahan M. Investigation of mechanical behavior of woven/knitted hybrid composites. Journal of the Textile Institute. 2017;**108**:1510-1517. DOI: 10.1080/00405000.2016.1258951

[6] Nesrin Sahbaz K, Yekta K, Huseyin O, Gokce O. Textile reinforced structural composites for advanced applications. Textiles for Advanced Applications. 2016:13. DOI: 10.5772/intechopen.68245

[7] Muralidhar BA. Tensile and compressive behaviour of multilayer flax-rib knitted preform reinforced epoxy composites. Materials and Design. 2013;**49**:400-405. DOI: 10.1016/j.matdes.2012.12.040

[8] Pei X, Shang B, Chen L, Li J, Tang Y. Compression properties of multilayer-connected biaxial weft knitted carbon fiber fabric reinforced composites. Composites. Part B, Engineering. 2016;**91**:296-305. DOI: 10.1016/j.compositesb.2015.12.041

[9] Kim KY, Curiskis JI, Ye L, Fu SY. Mode-I interlaminar fracture behaviour of weft-knitted fabric reinforced composites. Composites. Part A, Applied Science and Manufacturing. 2005;**36**:954-964. DOI: 10.1016/j.compositesa.2004.12.004

[10] Falconnet D, Bourban PE, Pandita S, Månson JAE, Verpoest I. Fracture toughness of weft-knitted fabric composites. Composites. Part B, Engineering. 2002;**33**:579-588. DOI: 10.1016/S1359-8368(02)00053-7

[11] Dadej K, Bienias J, Surowska B. On the effect of glass and carbon fiber hybridization in fiber metal laminates: Analytical, numerical and experimental investigation. Composite Structures. 2019;**220**:250-260. DOI: 10.1016/j.compstruct.2019.03.051

[12] Tabrizi IE, Kefal A, Zanjani JSM, Akalin C, Yildiz M. Experimental and numerical investigation on fracture behavior of glass/carbon fiber hybrid composites using acoustic emission method and refined zigzag theory. Composite Structures. 2019;**223**:110971. DOI: 10.1016/j.compstruct.2019.110971

[13] Swolfs Y, Gorbatikh L, Verpoest I. Fibre hybridisation in polymer composites: A review. Composites. Part A, Applied Science and Manufacturing. 2014;**67**:181-200. DOI: 10.1016/j.compositesa.2014.08.027

[14] Wisnom MR, Czél G, Swolfs Y, Jalalvand M, Gorbatikh L, Verpoest I. Hybrid effects in thin ply carbon/glass

unidirectional laminates: Accurate experimental determination and prediction. Composites. Part A, Applied Science and Manufacturing. 2016;**88**: 131-139. DOI: 10.1016/j.compositesa.2016.04.014

[15] Dong C, Davies IJ. Flexural and tensile strengths of unidirectional hybrid epoxy composites reinforced by S-2 glass and T700S carbon fibres. Materials and Design. 2014;**54**:955-966. DOI: 10.1016/j.matdes.2013.08.087

[16] Naito K, Oguma H. Tensile properties of novel carbon/glass hybrid thermoplastic composite rods. Composite Structures. 2017;**161**:23-31. DOI: 10.1016/j.compstruct.2016.11.042

[17] Greenhalgh E, Hiley CM. The influence of fibre architecture in the failure of polymer composites. In: Failure Analysis and Fractography of Polymer Composites, Woodhead Publishing Series in Composites Science and Engineering. 2009. pp. 279-355. DOI: 10.1533/9781845696818.279

[18] Zhao Y, Cao M, Lum WP, Tan VBC, Tay TE. Interlaminar fracture toughness of hybrid woven carbon-Dyneema composites. Composites. Part A, Applied Science and Manufacturing. 2018;**114**:377-387. DOI: 10.1016/j.compositesa.2018.08.035

[19] Bieniaś J, Dadej K, Surowska B. Interlaminar fracture toughness of glass and carbon reinforced multidirectional fiber metal laminates. Engineering Fracture Mechanics. 2017;**175**:127-145. DOI: 10.1016/j.engfracmech.2017.02.007

[20] Jung H, Kim Y. Mode I fracture toughness of carbon-glass/epoxy interply hybrid composites. Journal of Mechanical Science and Technology. 2015;**29**:1955-1962. DOI: 10.1007/s12206-015-0416-3

[21] Saidane EH, Scida D, Pac MJ, Ayad R. Mode-I interlaminar fracture toughness of flax, glass and hybrid flax-glass fibre woven composites: Failure mechanism evaluation using acoustic emission analysis. Polymer Testing. 2019;**75**:246-253. DOI: 10.1016/j.polymertesting.2019.02.022

[22] Swolfs Y, Geboes Y, Gorbatikh L, Pinho ST. The importance of translaminar fracture toughness for the penetration impact behaviour of woven carbon/glass hybrid composites. Composites. Part A, Applied Science and Manufacturing. 2017;**103**:1-8. DOI: 10.1016/j.compositesa.2017.09.009

[23] Yeter E, Erkliğ A, Bulut M. Hybridization effects on the buckling behavior of laminated composite plates. Composite Structures. 2014;**118**:19-27. DOI: 10.1016/j.compstruct.2014.07.020

[24] Nisini E, Santulli C, Liverani A. Mechanical and impact characterization of hybrid composite laminates with carbon, basalt and flax fibres. Composites. Part B, Engineering. 2017; **127**:92-99. DOI: 10.1016/j.compositesb.2016.06.071

[25] Bulut M, Erkliğ A, Yeter E. Hybridization effects on quasi-static penetration resistance in fiber reinforced hybrid composite laminates. Composites. Part B, Engineering. 2016; **98**:9-22. DOI: 10.1016/j.compositesb.2016.05.025

[26] ASTM International. ASTM D 5054 standard test methods for plane-strain fracture toughness and strain energy release rate of plastic materials. ASTM B Stand. 2013;**99**:1-9. DOI: 10.1520/D5045-14.priate

[27] Khan Z, Yousif BF, Islam M. Fracture behaviour of bamboo fiber reinforced epoxy composites. Composites. Part B, Engineering. 2017; **116**:186-199. DOI: 10.1016/j.compositesb.2017.02.015

[28] Silva RV, Spinelli D, Bose Filho WW, Claro Neto S, Chierice GO,

Tarpani JR. Fracture toughness of natural fibers/castor oil polyurethane composites. Composites Science and Technology. 2006;**66**:1328-1335. DOI: 10.1016/j.compscitech.2005.10.012

[29] Hanif WYW, Risby MS, Noor MM. Influence of carbon nanotube inclusion on the fracture toughness and ballistic resistance of twaron/epoxy composite panels. Procedia Engineering. 2015;**114**: 118-123. DOI: 10.1016/j.proeng.2015.08.049

[30] Kaman MO. Effect of fiber orientation on fracture toughness of laminated composite plates [0°/θ°]s. Engineering Fracture Mechanics. 2011; **78**:2521-2534. DOI: 10.1016/j.engfracmech.2011.06.005

[31] Arasan Ş, Aktaş M, Balcıoğlu HE. Fracture toughness of woven glass and carbon reinforced hybrid and non-hybrid composite plates. Polymer Composites. 2018;**39**:783-793. DOI: 10.1002/pc.23999

[32] Choupani N. Experimental and numerical investigation of the mixed-mode delamination in Arcan laminated specimens. Materials Science and Engineering A. 2008;**478**:229-242. DOI: 10.1016/j.msea.2007.05.103

[33] Hasanpour R, Choupani N. Rock fracture characterization using the modified Arcan test specimen. International Journal of Rock Mechanics and Mining Sciences. 2009;**46**:346-354. DOI: 10.1016/j.ijrmms.2008.07.004

[34] Yoon SH, Hong CS. Interlaminar fracture toughness of graphite/epoxy composite under mixed-mode deformations. Experimental Mechanics. 1990;**30**:234-239. DOI: 10.1007/BF02322816

[35] Jamali J, Mourad AHI, Fan Y, Wood JT. Through-thickness fracture behavior of unidirectional glass fibers/epoxy composites under various in-plane loading using the CTS test. Engineering Fracture Mechanics. 2016; **156**:83-95. DOI: 10.1016/j.engfracmech.2016.01.016

[36] Jamali J, Fan Y, Wood JT. The mixed-mode fracture behavior of epoxy by the compact tension shear test. International Journal of Adhesion and Adhesives. 2015;**63**:79-86. DOI: 10.1016/j.ijadhadh.2015.08.006

[37] Nikbakht M, Choupani N. Numerical investigation of delamination in carbon-epoxy composite using arcan specimen. International Journal of Mechanical and Mechatronics Engineering. 2008;**2**:750-757

[38] Okada H, Kadowaki S, Suzuki M, Yusa Y. J-integral computation for elastic-plastic materials with spatially varying mechanical properties. Engineering Fracture Mechanics. 2019; **207**:181-202. DOI: 10.1016/j.engfracmech.2018.12.029

[39] Tafazzolimoghaddam B, Curiel-Sosa JL. On the calculation of energy release rates in composite laminates by finite elements , boundary elements and analytical methods. Composites: Mechanics, Computations, Applications: An International Journal. 2015;**6**: 219-237. DOI: 10.1615/CompMechComputApplIntJ.v6.i3.40

[40] Gonzáles GLG, González JAO, Castro JTP, Freire JLF. A J-integral approach using digital image correlation for evaluating stress intensity factors in fatigue cracks with closure effects. Theoretical and Applied Fracture Mechanics. 2017;**90**:14-21. DOI: 10.1016/j.tafmec.2017.02.008

[41] Crump T, Ferté G, Jivkov A, Mummery P, Tran VX. Dynamic fracture analysis by explicit solid dynamics and implicit crack propagation. International Journal of Solids and Structures. 2017;**110–111**:

113-126. DOI: 10.1016/j.ijsolstr.2017.01.035

[42] Chabchoub M, Bouscarrat D, Vieille B, Gautrelet C, Beyaoui M, Taktak M, et al. Investigations on the mode I translaminar failure and determination of fracture toughness in woven-ply carbon fibers thermoplastic composites at high temperatures. Applied Acoustics. 2017;**128**:55-63. DOI: 10.1016/j.apacoust.2017.01.028

[43] Castrodeza EM, Perez Ipina JE, Bastian FL. Fracture toughness evaluation of unidirectional fibre metal laminates using traditional CTOD (δ) and Schwalbe (δ5) methodologies. Engineering Fracture Mechanics. 2004; **71**:1107-1118. DOI: 10.1016/S0013-7944(03)00138-3

[44] ASTM International. Designation: D 3039/D 3039M–00 Standard Test Method for Tensile Properties of Polymer Matrix Composite Materials 1. 2002. p. 15. DOI: 10.1520/D3039_D3039M-08

[45] ASTM International. Designation: D 3518/D 3518M–94 Standard Test Method for In-Plane Shear Response of Polymer Matrix Composite Materials by Tensile Test of a ±45° Laminate. 2001;**5**: 1-7. DOI: 10.1520/D 3518/D_3518M-94

[46] ASTM International. ASTM-D3410 standard test method for compressive properties of polymer matrix composite materials with unsupported gage section by shear loading. Annual Book of ASTM Standards. 2003;**15**(03):1-16. DOI: 10.1520/D3410

[47] ASTM International. ASTM D5379: Standard Test Method for Shear Properties of Composite Materials by the V-Notched Beam Method. 1998. pp. 1-13

[48] Garcia-Manrique J, Camas D, Lopez-Crespo P, Gonzalez-Herrera A. Stress intensity factor analysis of through thickness effects. International Journal of Fatigue. 2013;**46**:58-66. DOI: 10.1016/j.ijfatigue.2011.12.012

Chapter 5

Construction of an Artificial Neural Network-Based Method to Detect Structural Damage

Francisco Casanova-del-Angel, Daniel Hernández-Galicia and Xochicale-Rojas Hugo Alberto

Abstract

This chapter shows the framework used to obtain data with which the artificial neural network (ANN) was developed. It describes its geometry, properties of the material, sections of structural elements, and loads used. Then, the numerical model of the framework under study is developed in structural analysis using SAP2000® software in order to obtain its modal parameters. In addition, a program made in MATLAB® is shown, from which data with and without damage to the framework under study were obtained, and with which the ANN was developed. Data from the numerical model were used to corroborate data obtained with MATLAB®. The neural model used in this work to detect structural damage is described. Data on damage were obtained simulating a plastic hinge in various elements of a test framework, varying the position of the hinge. The above resulted in obtaining various damage conditions for the same framework, which data thus obtained were used to develop the network. Damage conditions were hierarchized based on their fundamental periods in order to know where is more damage, depending on location of the hinge within the framework. Upon completion of the research, we have concluded that the methodology implemented to detect structural damage is rather simple. It was carried out in four steps.

Keywords: damage detection, neural network, failure condition, framework, structural analysis

1. Introduction

Throughout their useful life, structures accumulate gradual damage as time passes. Such damage is caused by actions and natural phenomena such as earthquakes, winds, and explosions, among others. Therefore, and taking into account the functioning and the safety of structures, it is of paramount importance to detect damage, to follow up and monitor structures in order to know their physical conditions, and to increase safety and structural reliability. If damage is detected in a structure early, its evolution may be observed regarding magnitude and size, meaning it can be treated properly. Nowadays, there are various methods used to detect structural damage, all of which have advantages and disadvantages. There is no consensus that one method is the best. Therefore, for those who are not

IntechOpen

specialists on this topic, reading some specialized references would be useful [1–3]. In this article, an artificial neural network (ANN) is developed based on MATLAB's toolboxes [4]. Also, a program with source code was developed using MATLAB® [5], with which data were obtained (condensed matrices and modal parameters (frequencies, periods, and vibration modes) to train the network and, finally, to obtain an ANN capable to detect structural damage.

Damage has been defined as loss of stiffness. In order to identify such loss, plastic hinges in structural elements of the structure under study were simulated to obtain the condensed stiffness matrix with and without damage. Then, dynamic parameters were calculated, as well as their dynamic response with and without damage, which were used to develop the ANN. In this work, stiffness matrices with and without damage, that is, location of damage, are known.

2. Background and approach to the problem

For those devoted to analyze and design structures, it is very important to follow them up in order to know their conditions, specifically their physical conditions, in order to detect if there is any damage. Such follow-up on what has happened provides certainty regarding good structural safety for the structure and its inhabitants.

Some damage detection methods are visual and experimental, such as acoustic methods and methods based on magnetic, radiographic, and thermal fields. These experimental methods required access to the part of the structure where damage is located. Due to this limitation, these experimental methods detect damage in visible parts of the structure. The need to develop damage detection methods that may be applied to complex structures has led to develop methods that acknowledge changes in modal properties of structures. Every damage detection method using changes in modal properties of the structure is based on modal parameters (fundamental frequencies, vibration modes, and structure's vibration period), which are functions of physical properties of the structure (stiffness, mass, and damping). Therefore, changes in physical properties of structures such as stiffness cause changes in modal properties.

In civil engineering, studies have been developed to detect structural damage. They have applied ANNs with good results. However, these are recent developments in the area, and for this reason, researchers consider them significant for data development.

Rumelhart and his research team rediscovered the backpropagation learning algorithm (backpropagation). From 1986, the picture was encouraging with respect to the research and development of artificial neural networks [6, 7]. Wu et al. used a backpropagation neural network to identify damage in a three-stored building, modeled as a shear building in two dimensions, when being subject to earthquake excitation [8]. Damage was modeled, reducing stiffness of elements from 50 to 75%. Neural network was used to identify acceleration records and damage level of every element based on Fourier transform. The first 200 points in the fast Fourier transform (FFT) (0–20 Hz) were used as input in the network. A 200-node input layer and a 10-node hidden layer were selected as architecture of the network, and 42 training cases were used to train the network. In the first test, only acceleration data from the last floor were used, and the neural network was capable of identifying only the damage from the third floor with some exactness. The second network used acceleration data from two levels as input. This network was capable of diagnosing damage on the first and third levels with up to 25% approximation, but it was still incapable of predicting, with much precision, damage on the second level. The method is based on full knowledge of the acceleration record of two from three degrees of freedom.

Chen and Kim used a backpropagation neural network as a data processing technique for a metallic structure of a tridimensional bridge instrumented with accelerometers and strain gauges [9]. They studied feasibility to identify and locate structural damage using MATLAB's Neural Network Toolbox®. Vibration signals were obtained from a series of experiments on the scale structure. In their study, they developed four networks: the first one to identify damage using acceleration records, the second one to locate damage using data from strain gauges, the third one using signals simulated through analysis of finite elements and signals from the accelerometer, and the fourth one using acceleration records obtained from analysis of the finite elements model.

Marwala shows a technique based on using a set of neural networks with modal properties (frequencies and vibration modes) and the wavelet transform simultaneously in order to identify damage in structures [10]. The proposed method was tested with simulated data from a three-degrees-of-freedom spring-mass system. The result showed that performance of the method is not influenced by noise in data. Finally, the method was used to identify damage in 10 cylindrical shells. The set was capable to identify damage better than the three methods used separately.

In this work, artificial neural networks are applied to detect structural damage in steel buildings. An artificial neural network is designed to detect damage caused by actions present in the useful life of structures. In order to determine performance of the neural network, a structure is analyzed in various single and multiple damage scenarios. Damage is considered as loss of stiffness [11, 12]. The ANN is compared to data obtained in the numerical model and the program developed in MATLAB®. Learning of the developed neural network is based on epochs. The main idea here is that the network evolves for a period of time (called epoch), with an example sequence, under which it must render proper response. Upon completion of an epoch, the network is restarted in order that the new starting condition of such does not depend on the final condition of the previous epoch. Variation of weights is made upon completion of each epoch, and, as for the multilayer case, it is carried out by lots, that is, all the increases of patterns-epoch are calculated in an incremental manner, which is the one most used for recurrent networks.

As for hypotheses and objectives, it may be considered that side displacements were taken into account in order to make static condensation to horizontal displacements, that is, a stiff floor framework was considered. Only beams are axially stiff, and columns are completely flexible. Another hypothesis is that nodes would be absolutely stiff. In order to obtain data with and without damage (global stiffness matrices, condensed stiffness matrices, modal parameters, and dynamic response), plastic hinges (damage) were simulated on structural elements, based on the hypothesis that these hinges were near to the node, after the structure was subject to stress causing one or more elements to surpass their yield stress.

The objective is, throughout, to detect structural damage in steel buildings applying an ANN to monitor the structure in order to know and obtain proper knowledge of the physical conditions of such to be able to make reinforcement or rehabilitation decisions.

3. Method

A three-level framework is modeled in the structural analysis program SAP2000® [13]. From this modal parameter (frequencies, periods, and mode shapes) of the framework was obtained with and without damage. A program was developed in MATLAB® [5], from which the global and condensed stiffness matrices were obtained at horizontal degrees of freedom, with and without damage. Modal parameters and dynamic response with and without damage were calculated, which served for the development of the ANN, and a failure condition was considered to define a serious damage condition.

Data with damage were obtained by simulating a plastic hinge in various elements of the framework. The position of the hinges varied to obtain various damage conditions for the same framework. Data obtained were used to develop the network, and an ANN was obtained to detect structural damage. Regarding how damage conditions are determined, it must be considered that, in the moderate seism-resistant design of reinforced concrete frameworks, there is the fundamental philosophy to dimension and detail the elements of concrete frame in order that a plastic strong column-weak beam mechanism may be developed [14]. In addition, due to the complexity of the problem and for a quick and easy obtaining of data to train the ANN and for simplicity purposes, it was proposed that hinges or damages on the framework appear on the ends of structural elements (beams or columns). As for the question: Are all the damage conditions considered as possible? The answer is yes, since forming hinges in concrete frameworks usually takes place on the ends of their structural elements (beams or columns) [15]. It must be remembered that the ANN shown is in the development phase and still may not be compared to a real building. The ANN was trained for a three-story frame.

Damage detection was carried out in four steps: first, extraction of modal parameters and condensed matrix; second, establishing failure condition for a serious damage condition; third, treatment of modal data to be used in the development of the ANN; and fourth, detection of damage to be carried out with the ANN.

3.1 Static condensation

In order to carry out static condensation, the side stiffness matrix K_L was considered. It is associated to side coordinates of the floor, since seismic analysis of flat frameworks includes a single degree of freedom per floor. This is a stiff floor model, which only works for the analysis in view of the horizontal component of soil movement. Since the nodes are considered to be stiff, only the beams are axially stiff and the columns are flexible; condensation is carried out as follows: the original structure has n degrees of freedom (**Figure 1**). The condensed stiffness matrix is obtained first, at the desired degrees of freedom, GL. In our case, horizontal GLs are required, which are associated to displacement vector $\hat{u}$; in this case, the number of levels is identified as p, and K_L is the condensed matrix of p^*p, thus reducing the total stiffness matrix to condensed matrix $K_L = k_{11} - k_{12}k_{22}^{-1}k_{21}$.

3.2 Obtaining modal parameters

To obtain modal parameters, the diagonal mass matrix is $M = [m_1\, m_2\, m_3]$. Once the condensed stiffness matrix K_L and the mass matrix M have been obtained, the problem of eigenvalues and eigenvectors is solved: $(K_L - \omega^2 M)\Phi = 0$, with nonsingular M and symmetric and positive defined K_L. Then, there are N real roots ω_i, being $\omega_1 \leq \ldots \leq \omega_N$, for which reason to every eigenvalue corresponds an eigenvector ϕ_i which is called own vibration modes $\Phi = (\varnothing_1, \ldots, \varnothing_n)$ where Φ is the modal matrix. To find the period, ratio $T = \frac{2\pi}{\omega} \forall \omega = \sqrt{\frac{K}{M}}$ has been used. Therefore.

DOI: http://dx.doi.org/10.5772/intechopen.87193

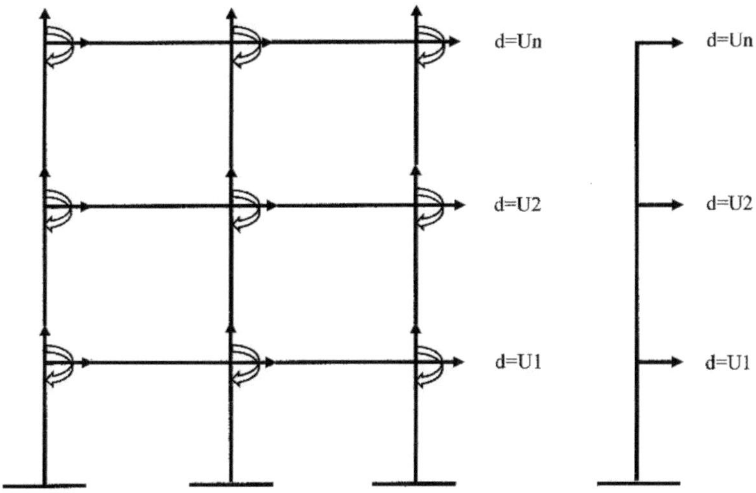

Figure 1.
n degrees of freedom system and condensed only at horizontal GL.

$$T = \begin{bmatrix} T_1 \\ T_2 \\ T_3 \end{bmatrix} \quad \omega = \begin{bmatrix} \omega_1 \\ \omega_2 \\ \omega_3 \end{bmatrix} = \begin{bmatrix} \omega_{11} & \omega_{21} & \omega_{N1} \\ \omega_{12} & \omega_{22} & \omega_{N2} \\ \omega_{N3} & \omega_{N3} & \omega_{NN} \end{bmatrix} \tag{1}$$

where T is the period and ω is frequency.

To obtain the dynamic response where there is movement on the base, the second-order linear differential equation $M\ddot{u} + C\dot{u} + K_L u = -mj\ddot{u}_g(t)$ is considered, where u is the vector of N horizontal displacements related to the movement of the soil, $\ddot{u}$ is acceleration, and $\dot{u}$ is speed. In addition, M, C, and K_L are the mass matrix, the damping matrix, and the side or condensed stiffness matrix, respectively. Every value of j influence vector is equal to one. The right side of the differential equation may be interpreted as the seismic forces $P(t) = -mj\ddot{u}_g(t)$ which is the function of time, t.

The distribution of these forces in height is defined by vector $s = mj$ and variation by time $\ddot{u}_g(t)$, which is acceleration of the ground. Such distribution of forces may be expressed as the summation of the distribution of modal inertial forces s_n; thus, $mj = \sum_{n=1}^{N} s_n = \sum_{n}^{N} \Gamma_n m \phi_n$, where ϕ_n is the ith vibration mode of the structure, Γ_n is the factor of modal participation, and m is the mass. Seismic forces may be expressed as $P(t) = \sum_{n=1}^{N} P(t)_n = \sum_{n=1}^{N} -s_n\ddot{u}_g(t)$.

The contribution of the ith mode to s is $s_n = \Gamma_n m \phi_n$ and to P(t) is $P(t) = -s_n\ddot{u}_g(t)$. Now, it may be seen that the response of the multiple degrees of freedom system $P(t)$ is completely in the ith mode, without any contribution from other modes. The equation representing the response of the system is $M\ddot{u} + C\dot{u} + K_L u = -s_n\ddot{u}_g(t)$. Using orthogonal properties of modes, it may be known that none of them, except the ith mode, contributes to the response. Therefore, the displacement of the floor is $u_n(t) = \phi_n q_n(t)$, where the modal coordinate $q_n(t)$ is governed by $\ddot{q}_n(t) + 2\zeta_n\omega_n q_n(t) + \omega^2{}_n q_n(t) = -\Gamma_n\ddot{u}_g(t)$. Consequently, $q_n(t)$ is the response of a single degree of freedom system. ζ_n is the damping coefficient, and ω_n is the vibration frequency. Therefore, floor displacements for multiple degrees of freedom system are given by $u_n(t) = \phi_n q_n(t)$.

4. Failure condition

The network will identify if there is damage or not. If there is any, it will consider it a failure condition, in order to determine if the structure has failed or not. Therefore, failure condition is defined as a limit service condition, that is, the presence of displacements affecting proper functioning of the building (**Figure 2**) not damaging its capacity to bear loads. In accordance with Article 209 of the Rules for Buildings in Mexico City [16, 17], admissible displacements are as shown in **Table 1**.

4.1 Presentation of a 2D framework

The flat framework of **Figure 3** was analyzed considering that beams are infinitely stiff regarding the columns and that the columns are fixed on the base. Stiffness of columns and beams and the spaces between galleries are in SI units. All the elements are made of steel with a I W14X311 section (see **Table 2**).

For the dynamic analysis of the framework, a program was developed in MATLAB®. With it condensed stiffness matrices and modal data (periods, frequencies, and modal shapes) were obtained. To obtain structural response, the framework was placed under stress by the seismic acceleration record in accordance with the department of communication and transportation for the seism that occurred in 1985 and a 5% damping. Data for analysis are as follows (see **Table 3**).

The geometry of the structure is described in **Table 4**. Column five of this table shows *Condition of the bar*, which may have four cases: (i) elements without hinges, (ii) elements with hinges *N* end, (iii) elements with hinges *F* end, and (iv) elements with both end hinges. For the example shown in **Table 4**, all the elements are in case i-elements without hinges. In addition, it must be mentioned that punctual

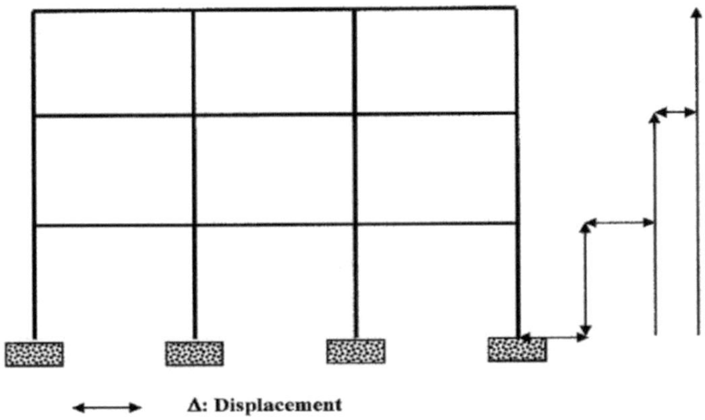

Figure 2.
Relative displacement of mezzanine.

For walls integrated into the framework	Δadm = 0.006 *h*
When walls are not integrated into the framework	Δadm = 0.012 *h*

Table 1.
Admissible values for horizontal displacements in accordance with RCDF [16], where h is the height of space between galleries under study.

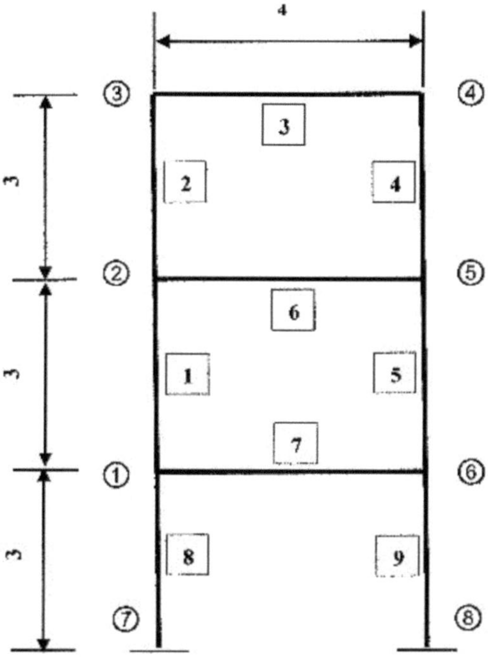

Figure 3.
Framework under analysis.

Beam I W 14X311	
Dimensions(m)	
t3	0.4348
t2	0.4122
Tf	0.0574
Tw	0.0358
t2b	0.4122
Tfb	0.0574

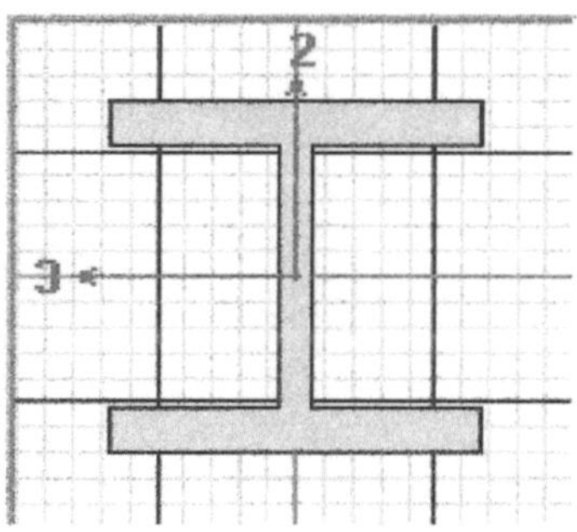

Table 2.
Dimensions of the section under study.

loads were placed on the nodes to simulate weights that would fall on the framework in service conditions. Magnitudes of loads were placed so that when carrying out the seismic analysis, the period of the framework was within the constant part of the seismic design spectrum of complementary technical standards [16, 17] and RCDF and has a behavior similar to that of most structures.

Loads applied on the framework are shown in **Figure 4**. Magnitude of loads was fixed at 25 ton in order that, when carrying out the seismic analysis, the vibration periods would fall on the constant part of the design spectrum of the NTC, Construction Rules for Mexico City.

Volumetric weight of steel	7.8 Ton/m^3
Number of degrees of freedom	24
Number of bars	9
Number of unknown displacements	18
Number of known forces	18

Table 3.
Degrees of freedom, number of bars and known/unknown forces on the framework.

Bar	L	I	E	Condition of the bar		A
	m	m^4	Kg/m^2			m^2
1	3	0.001802	20904200000	1	Element without hinge	0.059
2	3	0.001802	20904200000	1	Element without hinge	0.059
3	4	0.001802	20904200000	1	Element without hinge	0.059
4	3	0.001802	20904200000	1	Element without hinge	0.059
5	3	0.001802	20904200000	1	Element without hinge	0.059
6	4	0.001802	20904200000	1	Element without hinge	0.059
7	4	0.001802	20904200000	1	Element without hinge	0.059
8	3	0.001802	20904200000	1	Element without hinge	0.059
9	3	0.001802	20904200000	1	Element without hinge	0.059

Table 4.
Number of bars of the framework under study, lengths of elements, their geometric properties, and properties of material.

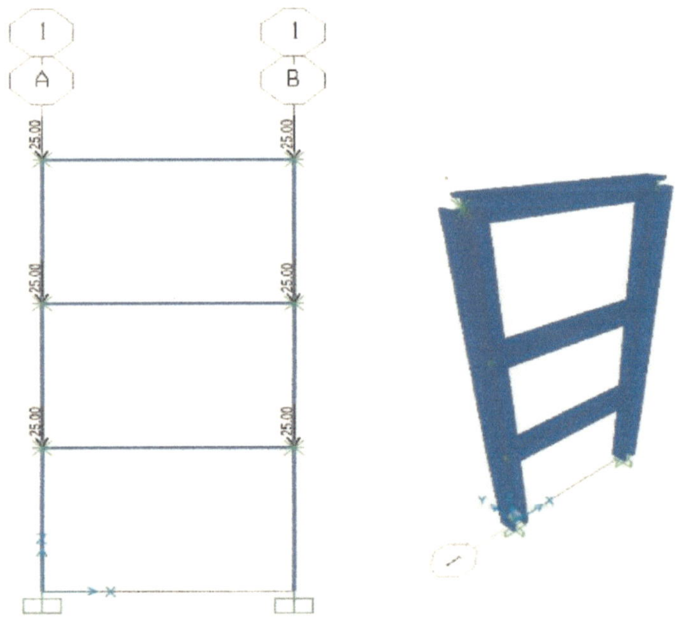

Figure 4.
Numerical model made in SAP2000® structural analysis software, with position of loads and their magnitudes.

4.2 2D numerical model

The framework shown in **Figure 3** was modeled with a structural analysis software, with the same geometry and properties of material, in order to obtain modal data (periods, frequencies, and modal shapes), as well as structural response for a I W14X311 section. The numerical model used was carried out to compare data obtained from such with those obtained with the program in MATLAB® in order to calibrate the numerical model.

4.3 Program in MATLAB® to simulate 2D framework

A program was developed in MATLAB®, with which the framework was analyzed, obtaining data with and without damage for the development of the ANN. For calculation of data with damage, the total stiffness matrix K_T was obtained from stiffness matrices with hinges on one or both ends. These conditions are defined in the program in order to carry out a structural and a dynamic analysis.

The program must be supplied with the geometry and properties of material, number of floors, damping, degrees of freedom to be condensed, NGLC, displaced degrees of freedom, NGLD, mass per floor, and accelerogram. The stiffness method is used to obtain the total stiffness matrix K_T. From this the reduced total stiffness matrix KTR, the reduced reordered total stiffness matrix KTRR, and finally, the

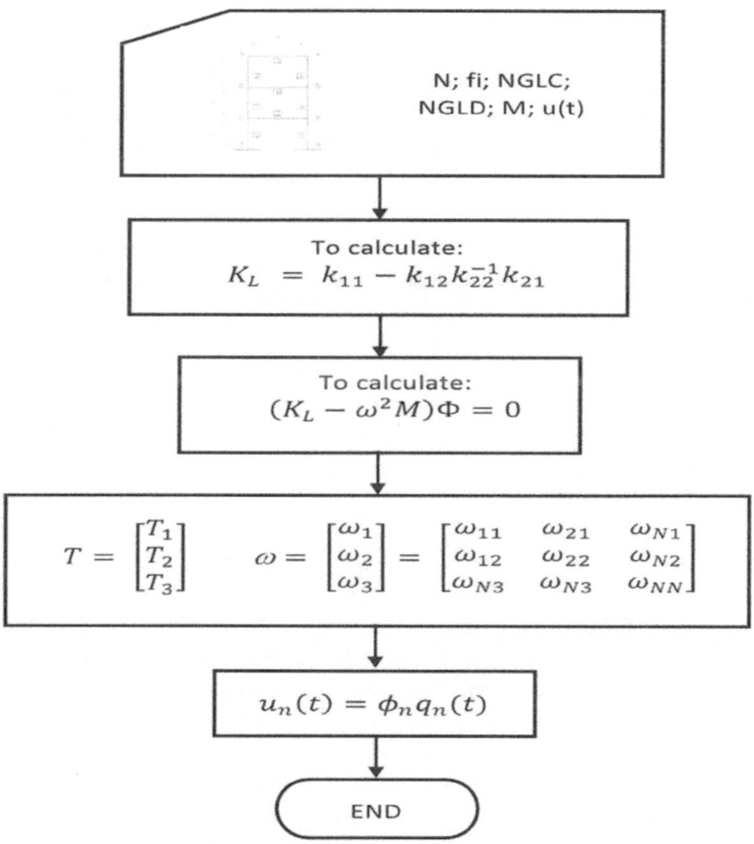

Figure 5.
Flowchart to obtain data with and without damage.

condensed matrix K_L are obtained. This is obtained to reduce the system in order to make a dynamic analysis to calculate horizontal displacements per floor, with and without damage. After that, the problem of eigenvalues and eigenvectors is solved to obtain frequencies and modal shapes of the framework. Finally, the response per floor is obtained applying an excitation on the base of the structure. The flowchart (**Figure** 5) shows the sequence of steps:

5. Application of the neural network to detect structural damage

Let us see three cases of application of the ANN which were created. A three-level framework was analyzed, in which plastic hinges on the ends of its elements were simulated. Plastic hinges varied in position to obtain various damage conditions. In addition, to simplify the problem, the n degrees of freedom system, GL, was reduced to a three-degrees-of-freedom system (**Figure 3**), from which 48 different damage conditions were obtained to train the network. In Case 1, 16 different conditions were used, from which only one is without damage while the remaining 15 have damage. For Case 2, 48 different conditions were used. For Case 3, the ANN was trained with modal matrices from 48 cases. The failure condition described was taken into account (to determine if the structure fails or not). For every state the condensed stiffness matrix K_L, its periods, frequencies, and modal matrices were obtained. These data were used to train the ANN.

5.1 Case 1: test of the ANN

For this example, the periods of every damage case were used to carry out the test of the ANN. From the above framework, a condition without damage was obtained, as well as 15 damage conditions. **Table 5** shows the elements damaged for every condition, as well as the location of damage. It should be mentioned that data were obtained with the program developed in MATLAB®. Structural analysis software SAP2000® was used only to verify that data obtained in the program created in MATLAB® were correct. **Figure 6** shows the structural response of Case 1 obtained with SAP2000® and the program created in MATLAB®, which were used to verify that results obtained with the program developed using MATLAB® were correct and, therefore, to train the ANN.

In addition, the network used is of the backpropagation type, which uses the error backpropagation algorithm or generalized delta rule (**Figure** 7). The network is provided with the following data to be trained: error = 0.001 and maximum number of epochs = 10,000. From the analysis of the framework, condensed stiffness matrices K_L were obtained for every condition. Since the framework has three levels, the condensed matrix K_L is 3×3. The principal diagonals were extracted to every K_L matrix to create a 3×16 matrix with the eigenvalues. This was made because the principal diagonal contains the stiffness coefficients to directions of displacements and spins, that is, coefficient K_{11} is the force generated in direction 1 when applying a unit displacement in the same direction. Also, a 3×16 matrix was created, with the periods obtained for every case. These two matrices were created in order to correlate them. Correlation between these two matrices created the weights matrix, to train the network. The matrix resulting from the correlation is 3×3. Once the weight matrix to start the network was obtained, the period's matrix was the input matrix for the network and, in turn, the objective matrix. This was made in order to verify that the network worked properly, that is, the output of the network should be the input period's matrix. With these last data, the network was trained. Obtaining data close to the periods wanted, **Table 6** shows the error obtained by the network, the number of epochs required to train the ANN, and the output weights.

Case	Damaged element								
	1	2	3	4	5	6	7	8	9
0	SD	SD	SD	SD	SD	SD	SD	SD	SD
1	SD	SD	D-EF	SD	SD	SD	SD	SD	SD
2	SD	SD	D-EN	SD	SD	SD	SD	SD	SD
3	SD	D-EF	SD	SD	SD	SD	SD	SD	SD
4	SD	SD	SD	D-EN	SD	SD	SD	SD	SD
5	SD	D-EN	SD	SD	SD	SD	SD	SD	SD
6	SD	SD	SD	D-EF	SD	SD	SD	SD	SD
7	SD	SD	SD	SD	SD	D-EF	SD	SD	SD
8	SD	SD	SD	SD	SD	D-EN	SD	SD	SD
9	D-EF	SD	SD	SD	SD	SD	SD	SD	SD
10	SD	SD	SD	SD	D-EN	SD	SD	SD	SD
11	D-EN	SD	SD	SD	SD	SD	SD	SD	SD
12	SD	SD	SD	SD	D-EF	SD	SD	SD	SD
13	SD	SD	SD	SD	SD	SD	D-EF	SD	SD
14	SD	SD	SD	SD	SD	SD	D-EN	SD	SD
15	SD	SD	SD	SD	SD	SD	SD	D-EN	SD

D, damage; SD, without damage. D-EN damage in extreme N (or mode vibration 1), and D-EF damage in extreme F (mode vibration 2).

Table 5.
Damaged elements and corresponding case.

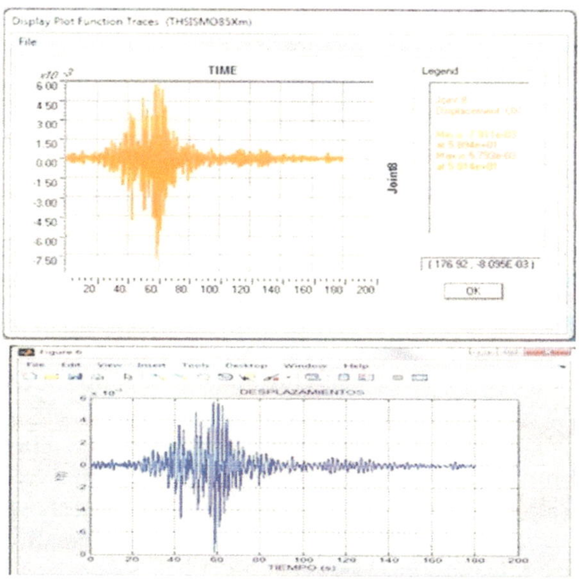

Maximum displacements m.

0.005793
-0.007811

Periods
0.318279
0.104332
0.051539

Maximum displacements m.

0.005787
-0.007804

Periods
0.318255
0.104709
0.052938

Figure 6.
The structural response obtained with SAP2000® and MATLAB® from Case 1, with maximum displacements and periods is shown.

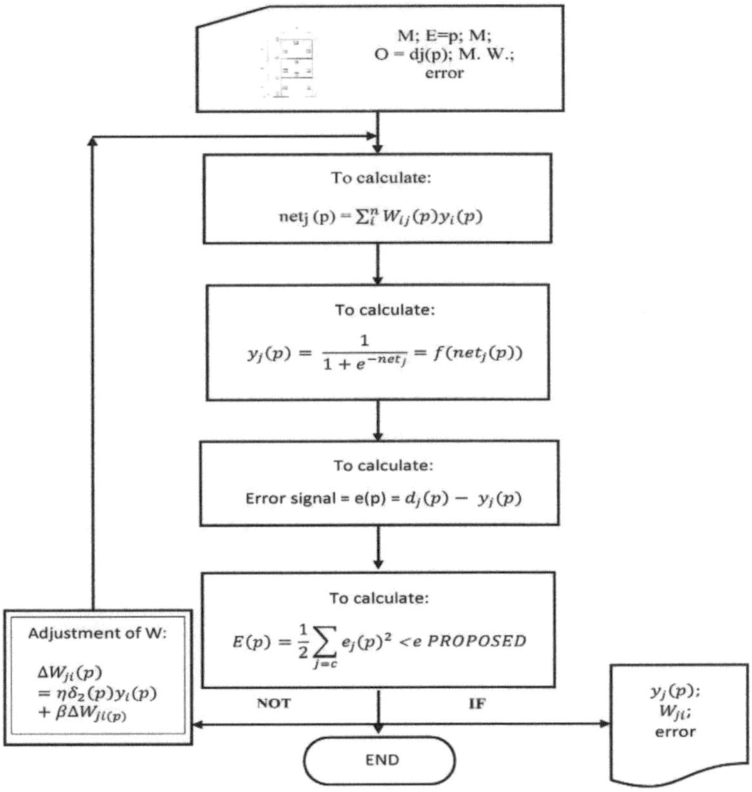

Figure 7.
Flowchart of the error backpropagation algorithm or the generalized delta rule for the case under study.

	Output weights			Average error
1	3.47499926	0.08577985	-0.88603064	1.048724
2	-0.43881538	-0.17416166	-0.88476444	5.329752
3	0.41886369	-0.85005851	0.22275651	5.459212

Table 6.
Output weights and average error between objective and calculated matrices.

5.1.1 Result

Results obtained where as follows:

It may be seen that the maximum displacements obtained with SAP2000® are almost equal to those obtained with MATLAB®, as well as the periods.

Inputs are the weight and period matrices and the objective. The output is a modified weight matrix and a calculated matrix close to the objective matrix. This diagram describes the error backpropagation algorithm or generalized delta rule explained above.

In tables in Annex I, it can be seen that the ANN works properly with a small error. In order to reduce the error obtained by the network, the ANN may be provided with a smaller error than that proposed, to reduce the difference between the required input and output. Here, the error of the network is 0.00099999 and the number of epochs required is 5976. **Table 6** for weights shows learning of the network.

Since the error of the network is lower than the error proposed, training is completed. In addition, the number of epochs required for training was 5976 of the 10,000 that were proposed at the beginning. As was previously mentioned, if the output calculated by the network is closer to the objective matrix given to it, a smaller error must be proposed. **Table 6** also shows the average error percentage between the objective and calculated matrices, which is acceptable. It must be considered that the error is a percentage measured between periods to be calculated by the ANN and the ones actually calculated.

5.2 Case 2: damage detection

In this application, condensed stiffness matrices were used to carry out all the training of the ANN, different from Case 1, where the periods were used for test purposes and to verify that the ANN was working properly. **Table 7** shows 48 damage conditions and which bar is damaged for every case.

5.3 Training of the network

To train the network, there are 48 condensed stiffness matrices, from which one belongs to the framework without damage matrix zero, and the remaining 47 are matrices with damage. Matrix number 47 was selected as the objective matrix, leaving it fixed for every training case. Matrix number 47 belongs to a framework with damage (**Figure 8**).

Figure 3 framework selected to train the ANN. Damage may be observed in bars 7 and 3. Framework number 47 was selected since it is known that an articulated column from the base of the framework is more unfavorable than an intermediate one, though numerically it was found that damage is higher if the hinge is in an intermediate column (**Figure 9**). Case 42 shown may be seen in **Figure 8**.

The input data to train the network were (i) condensed matrices for every framework; (ii) the weight matrix, created from the correlation of two matrices, the stiffness matrix containing the principal diagonal of each condensed matrix and the period matrix; and (iii) bias matrix, started with a unit matrix. The training began with condensed matrix number zero, the matrix without damage, and matrices were given in increasing order until matrix number 47. For the weight's matrix, the starting matrix was the one obtained in the correlation; the matrix obtained upon completion of training for each case became the input weight matrix for the following case and so on, until reaching matrix number 47. The same was carried out with the bias matrix. It should be mentioned that stiffness matrices were multiplied by a factor that reduced values to decimal numbers, because the network used only accepts values lower than the unit since they have a sigmoid activation function. In addition, training was supervised for every case. The maximum number of epochs (training cycles) proposed was 30,000, while the error calculated by the network should be lower than 0.001, and the η learning speed of 1 was proposed. It must be stressed that for each training case, and as was mentioned above, the objective matrix was matrix number 47 (damage case). This matrix was fixed and was the objective matrix for the 48 cases. For this purpose, the ANN had to find a weight matrix making the input matrix adjusted to the objective matrix, that is, it had to be equal or almost equal to the objective matrix. That means the ANN was working and learning properly.

5.3.1 Results

Table 8 shows the data proposed for training of Case Zero (Damage Zero) and data obtained by the ANN upon completion of training.

Case	Damaged element								
	1	2	3	4	5	6	7	8	9
0	SD	SD	SD	SD	SD	SD	SD	SD	SD
1	SD	SD	D-EF	SD	SD	SD	SD	SD	SD
2	SD	SD	D-EN	SD	SD	SD	SD	SD	SD
3	SD	D-EF	SD	SD	SD	SD	SD	SD	SD
4	SD	SD	SD	D-EN	SD	SD	SD	SD	SD
5	SD	D-EN	SD	SD	SD	SD	SD	SD	SD
6	SD	SD	SD	D-EF	SD	SD	SD	SD	SD
7	SD	SD	SD	SD	SD	D-EF	SD	SD	SD
8	SD	SD	SD	SD	SD	D-EN	SD	SD	SD
9	D-EF	SD	SD	SD	SD	SD	SD	SD	SD
10	SD	SD	SD	SD	D-EN	SD	SD	SD	SD
11	D-EN	SD	SD	SD	SD	SD	SD	SD	SD
12	SD	SD	SD	SD	D-EF	SD	SD	SD	SD
13	SD	SD	SD	SD	SD	SD	D-EF	SD	SD
14	SD	SD	SD	SD	SD	SD	D-EN	SD	SD
15	SD	SD	SD	SD	SD	SD	SD	D-EN	SD
16	SD	SD	SD	SD	SD	SD	SD	SD	D-EN
17	SD	SD	D-ENF	SD	SD	SD	SD	SD	SD
18	SD	D-EN	D-ENF	SD	SD	SD	SD	SD	SD
19	SD	SD	D-ENF	D-EF	SD	SD	SD	SD	SD
20	D-EF	SD	D-ENF	SD	SD	SD	SD	SD	SD
21	SD	SD	D-ENF	SD	D-EN	SD	SD	SD	SD
22	D-EN	SD	D-ENF	SD	SD	SD	SD	SD	SD
23	SD	SD	D-ENF	SD	D-EF	SD	SD	SD	SD
24	SD	SD	D-ENF	SD	SD	SD	SD	D-EN	SD
25	SD	SD	D-ENF	SD	SD	SD	SD	SD	D-EN
26	SD	SD	SD	SD	SD	D-ENF	SD	SD	SD
27	SD	SD	D-EF	SD	SD	D-ENF	SD	SD	SD
28	SD	SD	D-EN	SD	SD	D-ENF	SD	SD	SD
29	SD	D-EF	SD	SD	SD	D-ENF	SD	SD	SD
30	SD	SD	SD	D-EN	SD	D-ENF	SD	SD	SD
31	D-EN	SD	SD	SD	SD	D-ENF	SD	SD	SD
32	SD	SD	SD	SD	D-EF	D-ENF	SD	SD	SD
33	SD	SD	SD	SD	SD	D-ENF	SD	D-EN	SD
34	SD	SD	SD	SD	SD	D-ENF	SD	SD	D-EN
35	SD	SD	SD	SD	SD	SD	D-ENF	SD	SD
36	SD	SD	D-EF	SD	SD	SD	D-ENF	SD	SD
37	SD	SD	D-EN	SD	SD	SD	D-ENF	SD	SD
38	SD	D-EF	SD	SD	SD	SD	D-ENF	SD	SD
39	SD	SD	SD	D-EN	SD	SD	D-ENF	SD	SD
40	SD	D-EN	SD	SD	SD	SD	D-ENF	SD	SD
41	SD	SD	SD	D-EF	SD	SD	D-ENF	SD	SD
42	D-EF	SD	SD	SD	SD	SD	D-ENF	SD	SD
43	SD	SD	SD	SD	D-EN	SD	D-ENF	SD	SD
44	D-EN	SD	SD	SD	SD	SD	D-ENF	SD	SD
45	SD	SD	SD	SD	D-EF	SD	D-ENF	SD	SD
46	SD	SD	SD	SD	SD	SD	D-ENF	D-EN	SD
47	SD	SD	SD	SD	SD	SD	D-ENF	SD	D-EN

D, damage; SD, without damage; EN, end N; EF, end F; ENF, ends N and F.

Table 7.
Identification of damaged elements and in which case.

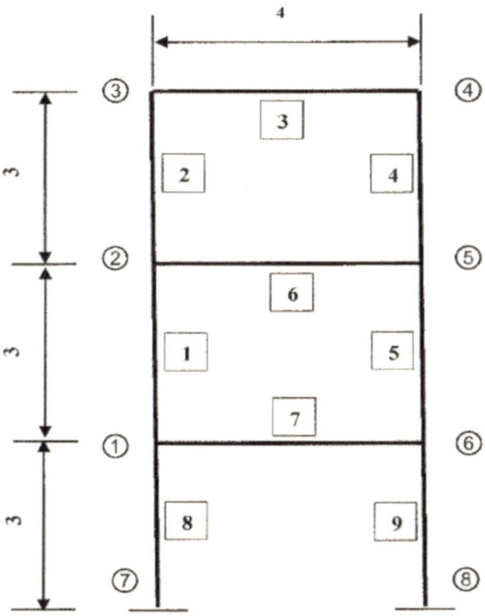

Figure 8.
Framework with the most serious damage (Case 42).

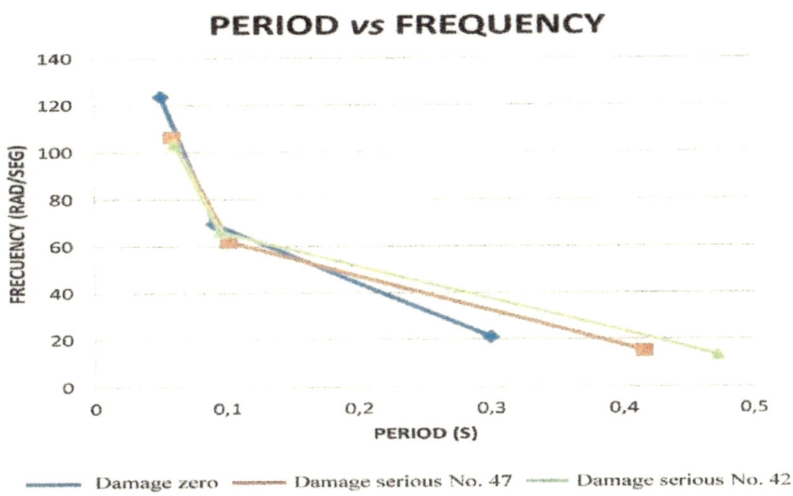

Figure 9.
The plot shows that damage is higher for Case 42 than for Case 47.

It can be observed that the 30,000 epochs proposed were required to train the ANN. In addition, it may be seen that the error calculated by the ANN is higher than that proposed as objective error. This error was left out, since it is not significant. The error calculated by the ANN could be reduced by increasing the number of epochs, but then training would be longer. Therefore, it was decided to leave these results out and, as it was already mentioned, the input matrix for the ANN, in this case the condensed matrix zero (Damage Zero), the objective matrix that was the condensed matrix of Case 47 (case with damage), and the output matrix calculated by the ANN. The ANN searches for a weight matrix to make the input matrix adjust to the objective matrix, which tells us how efficient its learning is. As may be seen in **Table 9**, the output matrix is almost equal to the objective matrix, with a 16% error.

Training data	
Start	10:05 a.m.
Lr	1
Error	0.001
Max. No. of epochs	30000

Data obtained	
Final	11:48 a.m.
Lr	1
Error red	0.16
No of epochs required	30000

Table 8.
Data proposed for training, calculated by the ANN for the Damage Zero case.

This procedure was made for the 48 cases considered, obtaining 48 matrices almost equal to the objective matrix (Case 47).

The dynamic properties of the framework were used to graph the period and the frequency, as well as to graphically observe the damage to the framework. **Figure 10** shows the case with Damage Zero, case with Damage 47, case with intermediate damage (Case 22), and the case with serious damage (Case 4) of 48 cases considered. These graphs show something already known, that is, that frequency is a function of stiffness and that period is a function of frequency. If stiffness changes, dynamic properties of the framework will also change. If stiffness decreases, frequency also decreases; therefore, the period increases. This shows that the structure lost stiffness (damage) and, therefore, became more flexible, thus increasing the vibration time (period). Therefore, the vibration period of Damage Zero framework is lower than the vibration period of the framework with serious damage (Case 42) (**Figure 10**). It should also be noted that mass was constant in every case. Graphs have been used to define an epsilon value (**Table 10**) both for the period and the frequency, respectively. With this parameter, through the ANN, the case of damage may be specified, as well as its range.

The damage conditions could also be placed hierarchically, based on their fundamental periods, in order to know where there is more damage, depending on the location of the hinge within the framework. It was observed that, when hinges were placed on the higher corners of the framework, damage was the same, regardless of the hinge being in the column or the girder. The same happened when the hinges were placed on the corners of the lower part of the framework. It should be highlighted that the damage is the same whether hinges are in the higher or lower corners of the framework. In other cases, the same did not happen. An important observed aspect is that cases where there was higher damage are those where hinges were inside the framework, that is, in intermediate elements. Case 47 was expected to be a serious

	Input matrix			Objective matrix			Output matrix		
	D0			D47			D47′		
	Condensed matrix Kg/m			Condensed matrix Kg/m			Condensed matrix Kg/m		
1	0.550475	-0.303859	0.059508	0.358309	-0.21264	0.050552	0.358558	0.000295	0.047987
2	-0.303859	0.439864	-0.205054	-0.21264	0.343081	-0.192423	0.000015	0.342884	0.000253
3	0.059508	-0.205054	0.151989	0.050552	-0.192423	0.150256	0.051943	0.000408	0.149131

Table 9.
The first table shows the input matrix, the second one shows the objective matrix, and the third one shows the output matrix calculated by the ANN.

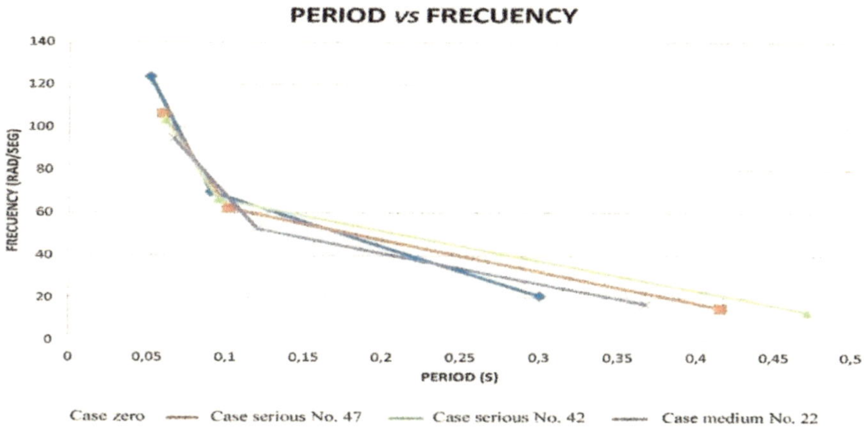

Figure 10.
This plot shows cases with Damage Zero, case with Damage 47, case with serious Damage 42, and case with intermediate Damage 22.

Period's epsilon			
Case 43	Highest period 0.472126659	Case 0	Lowest period 0.299830019
Difference	0.17229664	Seconds	
Frequency's epsilon			
Case 0	Highest frequency 20.95582465	Case 42	Lowest frequency 13.30826207
Difference	7.647562583	Rad/s	

Table 10.
Epsilon value for period and frequency.

damage case, where elements 7 and 9 of the frameworks are articulated. However, the serious damage case is Case 42, where hinges are in elements 1 and 7 of the frameworks.

5.4 Case 3: damage detection

In this example, the ANN was trained with modal matrices of every damage case. As it has already been mentioned, there are 48 different cases, from which one is the case without damage. For Case Zero, its condensed stiffness matrix K_L, its mass matrix M, its frequencies ω, and its modal matrix Φ are known (**Figure 3**).

There are also 47 cases of damage, through which the same information can be known.

$K_L{}^*, M^*, \omega^*, \Phi^*$ are, respectively, the condensed stiffness matrix with damage, the masses matrix with damage, and the modal matrix with damage (**Figure 11**). Since it is possible to know the dynamic properties of the framework $\omega^*; \Phi^*$ through an environmental vibration test, it was selected to train the ANN with the

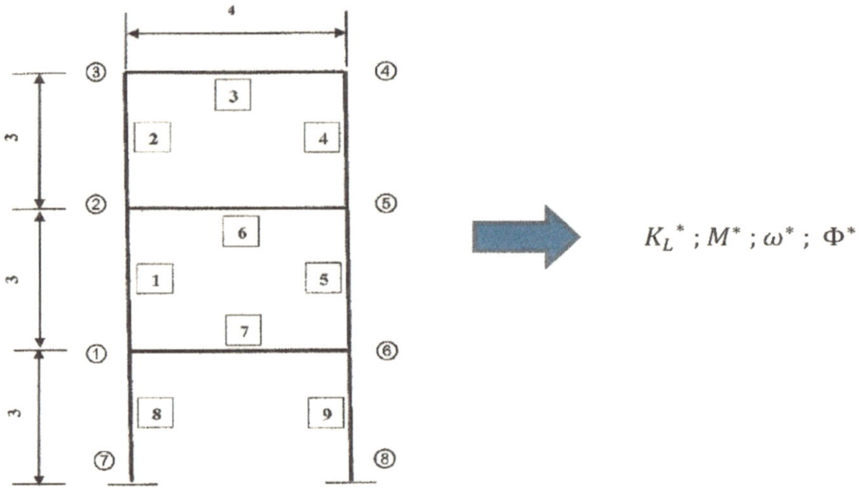

Figure 11.
Framework with damage, points represent hinges (damage).

modal matrices [18, 19]. With this, we aimed at the ANN defining matrix $K_L{}^*;M^*$, which would correspond to the said modal matrix. Once the ANN was trained and it provided searched for data for $K_L{}^*;M^*$, the failure condition was taken into account, in this case a service limit condition. Such service limit condition was defined as displacement of mezzanine $u^i{}_o$ and $u^i{}_j$ as well as differential displacements $\Delta^i{}_o$ y $\Delta^i{}_j$. Therefore, if

$$u^i{}_j > u^i{}_o \textit{ Structure fails}$$

$$\Delta^i{}_j > \Delta^i{}_o \textit{ Structure fails}$$

Here the subscript shows the case of damage, and the superscript shows the number of levels. Thus, $u^i{}_o$ shows the case of Damage Zero of level i. To obtain u_o, Δ_0, u_j, and Δ_j, the load condition was defined as the spectrum of design in Mexico City. For the calculation of u_o and Δ_0, the Damage Zero framework data were used K_L and M, but for the calculation of u_j and $\Delta_j K_L{}^*$ and M^*, the ANN gives as a result after having been trained were used, providing a modal matrix that may be obtained through an environmental vibration test.

6. Conclusions and discussion

According to Test 1, the ANN works properly calculating the objective matrix in an efficient manner, with an average 5% error. This may be lower if the number of training epochs of the ANN increases.

From Test 2 it can be concluded that, if training cases increase, the number of epochs required for learning of the ANN is higher. This is because input data of the ANN is larger, and therefore, more time is required for training. Also, from Test 2 it is concluded that the ANN has a 16% error when calculating the objective matrix, which corresponds to the increase of training cases. However, if the number of training epochs increases, the error of the ANN will decrease because the ANN will have more learning time.

Cases of serious damage are those where hinges are in intermediate elements, and cases where damage is of the same magnitude are those where hinges are in the higher and lower corners of the framework.

A hinge column on the base is more unfavorable than a hinge column in higher floors, but numerically, damage is higher when there are hinge elements in higher floors.

It may be interesting to see the damage level and period relation. How is period and damage related, the level of damage and period elongation is consistent. How much different damage levels coincide with the same period value and vice versa. But these points of interest may not be addressed in this work, since there was no parameter to measure the damage level on the framework. Damage is only found with changes in modal parameters, as well as with the increase of the vibration period.

Upon completion of the research, we have concluded that the methodology implemented to detect structural damage is rather simple. It was carried out in four steps: (i) extraction of modal parameters and condensed matrix, (ii) establishing failure condition for a serious damage condition, (iii) treatment of modal data to be used in the development of the ANN, and (iv) detection of damage to be carried out with the artificial neural network.

Acknowledgements

This article and its corresponding research were carried out, in part, with the research projects IPN SIP-20181253 and IPN SIP-20196119. The authors thank the reviewers for carefully reading the paper and for their constructive comments and suggestions which have improved this paper.

Annex I

Constructiong of an artificial neural network for detection of structural damage (**Tables I1–I4**).

		Principal stiffness kg/m		
Damage type	**Level**	**1**	**2**	**3**
SD0	*0*	55047511.35	43986421.33	15198916.05
D1	*1*	54877708.7	36097371.39	9399138.702
D2	*2*	54771544.01	35317166.31	9121527.737
D3	*3*	54771544.01	35317166.31	9121527.737
D4	*4*	54877708.7	36097371.39	9399138.702
D5	*5*	52859001.57	30346342.81	9268800.799
D6	*6*	53635426.09	32269306.5	9737254.051
D7	*7*	51836691.31	43952306.34	13441604.94
D8	*8*	51573961.66	43939537.94	13450263.27
D9	*9*	44998259.72	29296754.08	13808325.01

		Principal stiffness kg/m		
Damage type	Level	1	2	3
D10	*10*	46785291.09	31313923.98	14088832.2
D11	*11*	38598400.77	34400828.39	14998080.93
D12	*12*	41125388.66	35987149.14	15086548.41
D13	*13*	54,908,398	41044048.98	15126241.06
D14	*14*	54923344.34	40895188.06	15112225.05
D15	*15*	36667939.84	41919304.71	15169346.77

Table I1.
Matrix stiffness formed with the principal diagonal of each condensed matrix K_L.

		Periods		
Damage type	Level	1	2	3
SD0	*0*	0.299830	0.090200	0.050792
D1	*1*	0.318255	0.104710	0.052938
D2	*2*	0.318365	0.105415	0.053246
D3	*3*	0.318365	0.105415	0.053246
D4	*4*	0.318255	0.104710	0.052938
D5	*5*	0.303981	0.104247	0.055981
D6	*6*	0.303965	0.103230	0.054903
D7	*7*	0.348754	0.096427	0.051103
D8	*8*	0.348938	0.096669	0.051148
D9	*9*	0.323188	0.090897	0.060636
D10	*10*	0.323013	0.090790	0.058691
D11	*11*	0.319120	0.092516	0.060238
D12	*12*	0.318989	0.092179	0.058335
D13	*13*	0.362777	0.092909	0.051041
D14	*14*	0.362999	0.093013	0.051068
D15	*15*	0.311588	0.100941	0.055899

It is the matrix of entry to the network and target matrix.

Table I2.
Matrix of periods.

		Weights correlation K *vs.* P	
1	0.19607059	0.36461043	−0.75429818
2	0.47564924	−0.19211267	−0.73301055
3	0.39640392	−0.89248216	0.149395599

Table I3.
Input matrix to the network weights.

Periods, input matrix, and target matrix				
Damage type	**Level**	**1**	**2**	**3**
SD0	*0*	0.299830	0.090200	0.050792
D1	*1*	0.318255	0.104710	0.052938
D2	*2*	0.318365	0.105415	0.053246
D3	*3*	0.318365	0.105415	0.053246
D4	*4*	0.318255	0.104710	0.052938
D5	*5*	0.303981	0.104247	0.055981
D6	*6*	0.303965	0.103230	0.054903
D7	*7*	0.348754	0.096427	0.051103
D8	*8*	0.348938	0.096669	0.051148
D9	*9*	0.323188	0.090897	0.060636
D10	*10*	0.323013	0.090790	0.058691
D11	*11*	0.319120	0.092516	0.060238
D12	*12*	0.318989	0.092179	0.058335
D13	*13*	0.362777	0.092909	0.051041
D14	*14*	0.362999	0.093013	0.051068
D15	*15*	.0311588	0.100941	0.055899
Periods, output matrix				
Damage type	**Level**	**1**	**2**	**3**
SD0	*0*	0.30661	0.09916	0.05424
D1	*1*	0.32025	0.09805	0.05403
D2	*2*	0.32029	0.09801	0.05401
D3	*3*	0.32029	0.09801	0.05401
D4	*4*	0.32025	0.09805	0.05403
D5	*5*	0.30896	0.09837	0.05378
D6	*6*	0.30914	0.09847	0.05381
D7	*7*	0.34395	0.09714	0.05503
D8	*8*	0.34409	0.09713	0.05502
D9	*9*	0.32224	0.09747	0.05483
D10	*10*	0.32248	0.09763	0.05481
D11	*11*	0.31927	0.09763	0.05467
D12	*12*	0.31953	0.09779	0.05466
D13	*13*	0.35497	0.09666	0.05549
D14	*14*	0.35515	0.09665	0.05549
D15	*15*	0.31459	0.09814	0.05409
Difference, target matrix, and output matrix				
Damage type	**Level**	**1**	**2**	**3**
SD0	*0*	−0.006784	−0.008962	−0.003451
D1	*1*	−0.001994	0.006660	−0.001092
D2	*2*	−0.001921	0.007404	−0.000760

Difference, target matrix, and output matrix				
Damage type	Level	1	2	3
D3	*3*	−0.001921	0.007404	−0.000760
D4	*4*	−0.001994	0.006660	−0.001092
D5	*5*	−0.004983	0.005873	0.002200
D6	*6*	−0.005173	0.004756	0.001091
D7	*7*	0.004801	−0.3000715	−0.003928
D8	*8*	0.004845	−0.000458	−0.003877
D9	*9*	0.000947	−0.006573	0.005807
D10	*10*	0.000530	−0.006841	0.003883
D11	*11*	−0.000149	−0.005118	0.005573
D12	*12*	−0.000541	−0.005614	0.003680
D13	*13*	0.007803	−0.003753	−0.004452
D14	*14*	0.007851	−0.003637	−0.004426
D15	*15*	−0.003003	0.002805	0.001814

Table I4.
The ANN output data.

Author details

Francisco Casanova-del-Angel[1*], Daniel Hernández-Galicia[1]
and Xochicale-Rojas Hugo Alberto[2]

1 Instituto Politécnico Nacional, Mexico City, Mexico

2 Instituto Mexicano del Seguro Social, Mexico City, Mexico

*Address all correspondence to: fcasanova@ipn.mx

References

[1] Futao Z, Wu Y. A rapid structural damage detection method using integrated ANFIS and interval modeling technique. Applied Soft Computing. 2014;**25**:473-484. DOI: 10.1016/j.asoc.2014.08.043

[2] Rezvani K, Maia NMM, Sabour MH. A comparison of some methods for structural damage detection. ResearchGate. 2017;**42**(4):649-659. DOI: 10.24200/sci.2017.4494

[3] Ancona Lazcano AR, Salgado Estrada R, Zamora Castro SA, Marcial MF. Evaluación de Métodos de Detección de Daño en Estructuras mediante uso de Vibraciones. In: XVIII Congreso Nacional de Ingeniería Sísmica. Aguascalientes, Mexico: Sociedad Mexicana de Ingeniería Sísmica; 2011

[4] Demuth H, Beale M. Neuronal Network Toolbox. For Use with MATLAB. User's Guide. Version 4. The MathWorks Inc; 1992–2004

[5] Gilat A. Matlab una introducción Con Ejemplos prácticos. Barcelona, España: Editorial Reverté; 2006. ISBN: 8429150358, 9788429150353

[6] Rumelhart DE, Hinton GE, Williams RJ. Learning internal representations by error propagation. In: Parallel Distributed Processing: Explorations in the Microstructure of Cognition. Vol. I. Cambridge, MA, USA: MIT Press; 1986. pp. 318-362. ISBN: 0-262-68053-X

[7] Rumelhart DE, McClelland JL, The PDP Research Group. Parallel Distributed Processing Exploration in the Microstructure of Cognition. Volume 1: Foundation. Cambridge, MA: MIT Press; 1986

[8] Wu X, Ghaboussi J, Garrett JH. Use of neural networks in detection of structural damage. Computers and Structures. 1992;**42**(4):649-659. DOI: 10.1016/0045-7949(92)90132-5

[9] Chen SS, Kim S. Neural networks based sensor signal monitoring of instrumented structures. In: Proc. Struct. Congr. XIII. 1995

[10] Marwala T. Damage identification using committee of neural networks. Journal of Engineering Mechanics. 2000;**126**(1):43-50 ASCE. ISBN (Print): 0733-9399

[11] Doebling SW, Farrar CR, Prime MB, Shevitz DW. Damage Identification and Health Monitoring of Structural and Mechanical Systems from Changes in their Vibration Characteristics: A Literature Review, United States. 1996 DOI: 10.2172/249299

[12] Barrios R, Doz G, Iturrioz I. Detección de daño en estructuras utilizando propiedades dinámicas. In: Información Tecnológica, Centro de Información Tecnológica. Vol. 11. La Serena, Chile; 2000. pp. 117-122

[13] Computers and Structures Inc. Getting Started with SAP2000 Linear and Nonlinear Static and Dynamic Analysis and Design of Three-Dimensional Structures. Berkeley, California, USA; 2009

[14] Solís Ortiz J, Martinez A, de Castro SM. Implicaciones en el Diseño de Marcos de Concreto Reforzado de los Criterios Para Asegurar el Mecanismo Plástico Columna Fuerte-Viga débil. Puebla, México: Sociedad Mexicana de Ingeniería Estructural, A.C; 2002

[15] Ávila J, Meli R. Análisis de la Respuesta de Edificios típicos Ante el Sismo del 19 de Septiembre de 1985. Mexico City: Instituto de Ingeniería, UNAM; 1987

[16] Reglamento de Construcciones para el Distrito Federal (RCDF). Reglamento publicado en Gaceta Oficial del Distrito

Federal, el 29 de enero de 2004. Última Reforma Publicada en la Gaceta Oficial de la Ciudad de México; 15 de Diciembre de 2017

[17] Luis AS, Max BS. Reglamento de Construcciones Para el Distrito Federal. 2005. ISBN: 968-24-7188-5

[18] Çalık İ, Bayraktar A, Türker T, Ayan AO. Dynamic identification of historical Molla Siyah mosque before and after restoration. Research on Engineering Structures and Materials. 2017;**3**(3): 164-175. DOI: 10.17515/resm2016.82st0728

[19] Salgado R. Damage detection methods in bridges through vibration monitoring: Evaluation and application [tesis doctoral]. Guimaraes, Portugal: Universidad de Minho-ISISE; 2008. p. 320

Chapter 6

The Mechanism of Misalignment of Saw Cutting Crack of Concrete Pavement

Chatarina Niken

Abstract

Misalignment cracks are transverse cracks that occur not in the cutting line but that are shifted within less than 500 mm of the cutting line. This crack does not cross other segments. This paper describes the mechanism of the formation of misalignment cracks and the stresses that occur in concrete pavement under plastic and brittle condition. This paper was written based on observations of misalignment cracks on toll roads in Lampung Province, Indonesia. Bending strength of the concrete pavement is ±4.5 MPa. This crack was found at the concrete age of 18–72 hours. This research is supported by observing deformation and inner temperature in the laboratory on a concrete plate with compressive strength of 60 MPa measuring 300 cm × 160 cm × 15 cm, which is placed on several supports. Observations were made every 15 minutes for 90 days. Misalignment cracks occur because cutting in concrete pavement is done in brittle conditions. Misalignment cracks were also found on one side, which dowel shift. In this phenomenon, misalignment cracks follow dowel shifting.

Keywords: concrete, crack, deformation, pavement, cutting, shrinkage

1. Introduction

Concrete or rigid pavement is often chosen as a road material that must withstand heavy loads. To withstand heavy loads, high-quality concrete is needed, which can only be made with limited water. The behavior of high-strength concrete has been studied [1, 2]. Apart from its ability, concrete pavement does not require expensive maintenance. Even with the best slab design and proper construction, however, it is unrealistic to expect crack-free and curl-free floor [3]. Humans, nature, and the type of the structural element affect the performance of concrete pavement forever.

Expansion and shrinkage naturally occur in the hydration process, due to, among others, product hydration growth, pore pressure, disjoining pressure, and surface force. It's called microprestress. It is easy to understand that microprestress always changes during time. Shrinkage and expansion are two behaviors that occur alternately in the concrete pavement. Concrete must be able to withstand these changes so that performance is good. There are many uncertainties to predict the shrinkage and expansion such as concrete composition, aggregate, relative humidity, and element structure geometry and ratio between open surface area and volume. Base course type also influences the concrete strength [4].

The concrete pavement has large surface contact to environment. Environment affects product hydration quality and drying shrinkage. Without covering the pavement right after pouring, inner water evaporates. Wind also accelerates inner water evaporation. This condition leads to the hydration process that occurs under limited water. It makes concrete more sensitive for premature cracks. More relative humidity is correlated to lower shrinkage [5]. The weather almost always has a role in the occurrence of uncontrolled cracking [6].

Hydration process results in heat that can reach 60°C [7]. A lot of concrete pores make the surrounding temperature to penetrate the concrete. Pouring fresh concrete during high surrounding temperature leads to the increase of the inner temperature. Heat evolution at any given time provides the change of mineralogical composition of the system [8]. An increase in equivalent linear temperature gradient leads to the increase in curvature for the unrestrained slab that was 7% higher than the restrained [9]. High inner temperature makes the hydration process occur with limited water. The consequence of this is premature deterioration and the change of expansion and shrinkage at early-age occur with high fluctuation.

Shrinkage is the root cause of random cracking in unreinforced concrete floors. From shrinkage to temperature, a large tensile strength in transversal direction on the bridge was studied [10]. Slow rate of shrinkage is a factor difficulty to predict the shrinkage accurately from laboratory measurement; therefore, minimum variant coefficient is assumed to be 20% to predict long-term shrinkage. Besides, this road has a dynamic problem. Compression and tension occur when the road is operational. The effect of shrinkage and creep was inserted by a solution for dynamic problem in reinforced concrete plate and beam [11].

Before applied load, microprestress occur in the concrete. Cracks can appear if no space is available for expansion and shrink. Open freely and joints help to release stresses. Joints of reinforced isolation were not as effective in reducing the edge stresses as compared to thickened joints [12]. The joints also help control unsightly transversal and longitudinal cracking, and also random slab cracking [13]. Cracks in joints are the consequence of concrete behavior in the initial period of use of pavement structures, which can affect structural properties of concrete pavements [14]. An overview of stress prediction for cracking of jointed plain concrete pavement was done in 1925–2000 [15].

Construction joint or contraction joint (also called control joint) can be made by open freely. Open freely make concrete pavement as unrestrained condition. Without joints, most concrete pavement would come in contact with cracks within 1 or 2 years after placement [16], except for continuously reinforced concrete (CRC) pavement, which has no joints. The open freely to be a place releasing the energy to avoid the premature and mature cracks. Open freely should be made in rigid pavement by saw cutting. While determining the location of the saw cutting, the distance where the shrinkage and expansion are still within the limits of the ability of concrete pavement must be considered. The depth of the saw cutting must really make this place the weakest location. Saw cutting must also consider the maturity of the concrete to be able to withstand microprestress at that age. Without those three things, we just move the location of the transverse crack. Late or shallow saw cutting of longitudinal joints causes longitudinal cracking [17]. Crack will shift from saw cutting line, hereinafter referred to as misalignment crack. Initial crack is not caused by misalignment of concrete pavement [18]. Vibration in the construction of concrete pavement in open traffic conditions makes the concrete pavement surface more at risk for the onset of hair cracks [7].

The inability of the concrete pavement to withstand all of the stresses can cause defects such as transverse and longitudinal cracking, cracks over dowel bars, cracks over slab culvert, blow up, crocodile cracks, hair cracks, etc. Road foundation (base/

subbase) settlement can cause a break in the rigid pavement or appear as a corner break depending on the location of the settlement. The cause of the crack must be understood in order to obtain the best concrete pavement performance.

These cracks and distresses do indicate the failure of concrete slab, but more than that, they indicate the human failure to understand the few basic and fundamental things related to concrete material and pavement. Implementers of concrete roads must have a sufficient understanding of the mechanism of the cracks in the concrete pavement to avoid the cracks. When it comes to the repair of distressed concrete pavement, very limited options are available. It is very difficult and time- and money consuming to repair hardened concrete [18]. Based on this statement, and limited analysis of misalignment crack, the mechanism of cracking needs to be studied and understood.

2. Plate deformation

2.1 Laboratory test specimen

Deformation of plate was found by full scale test in laboratory. The research was conducted in Jakarta, Indonesia, with humid tropical weather. This research was performed experimentally using one specimen of 3.000 × 1.600 × 150 mm^3, with four embedded vibrating wire strain gauges (EVWSG) as shown in **Figure 1a**. The plate was supported by many cylinders with 15 diameter and 30 cm high in horizontal position as shown in **Figure 1b**.

High-performance concrete with target compressive strength of 60 MPa and slump flow diameter of 35 ± 2 cm was applied. The plate was casted at 22.10 the night with surrounding temperature of 25.4°C and 94% relative humidity.

2.2 Laboratory observation of plate

In laboratory observation, shrinkage was measured as strain changes against time by installing four embedded vibrating wire strain gauge (EVWSG) in the specimen (**Figure 1a**). The EVWSG has abilities to detect the strain up to 3000 με with accuracy of about 0.025% and concrete temperature between −80°C and 60°C with about 0.5% accuracy. The specimens were covered with a plastic sheet to reduce evaporation of water immediately after casting. Observation was performed right after pouring as follows: (a). 0–24 hours, every 15 minutes; (b) 24–48 hours, every 60 minutes; and (c) 48–72 hours, every 2 hours. Relationship between strain and time and inner temperature and time were obtained.

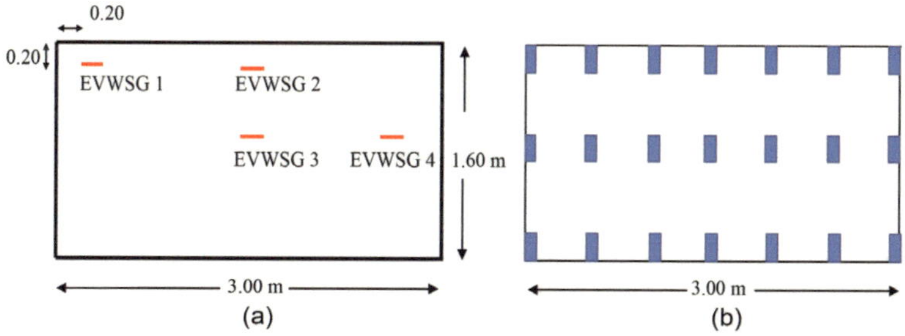

Figure 1.
The scheme of plate: (a) the placement of embedded vibrating wire strain gauges (EVWSG), (b) placement of plate supporter.

2.3 Relationship between deformation and time

Relationship between deformation of every position as shown in **Figure 1** during the first 90 hours was displayed in **Figure 2**.

Figure 2c shows that the highest deformation occurs in the middle of the plate, while the next sequence is in the corner of the plate (**Figure 2a**). The middle side of the elongated and shortened plates shows smaller deformation. Final concrete shrinkage in structure is about 600×10^{-6} [19]. Concrete capability to sustain tensile strain is about 600×10^{-6}. Cracks will appear if the shrinkage was restrained. Slab ≤4.5 m length will reduce tension stress. Slab with fc '60 MPa can shrink ±0.00022 mm/mm in the first 24 hours and at 1760 days of ±0.0003 mm/mm. Open freely is needed so that shrinkage can occur without obstacles. If there is no open freely, cracks will occur even though high-shrinkage cement is used.

2.4 Relationship between inner temperature and time

Inner temperature in early age shows a sharp increase in the age of 5–12 hours and then decreases gradually (**Figure 3**). The heat in concrete shows the activeness of the hydration process. The middle part of the plate shows the highest temperature, which is around 58°C (**Figure 3c**), while the angle is slightly below (**Figure 3a**). The middle side of the elongated and shortened side shows the peak of the lowest inner temperature, but it is still more than 50°C even though the concrete pavement is already 90 hours old and curing the humidity at the age of 24–168 hours (**Figure 3b** and **d**). At the age of 12–30 hours, there is a fairly large inner temperature gradient of ±28°C. Open freely is needed to release its energy. Open freely must be available before the concrete can hold the shrinkage energy. If there is a delay, a crack will occur. When curing the humidity is done, the inner temperature goes up to stabilize gradually until around 30° Celsius (**Figures 2** and **3**). It approaches the ambient air temperature, so the risk of cracking due to temperature difference decreases.

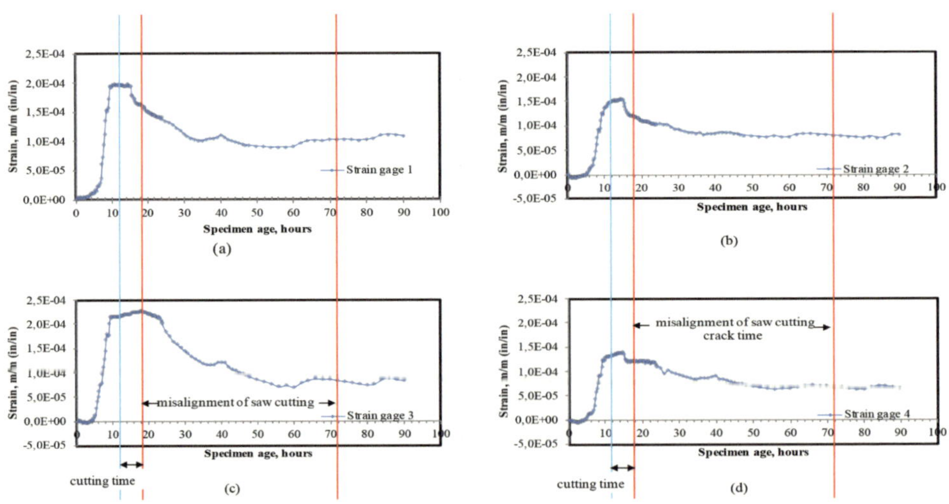

Figure 2.
Relationship between deformation and time of slab: (a) EVWSG 1, (b) EVWSG 2, (c) EVWSG 3, and (d) EVWSG 4.

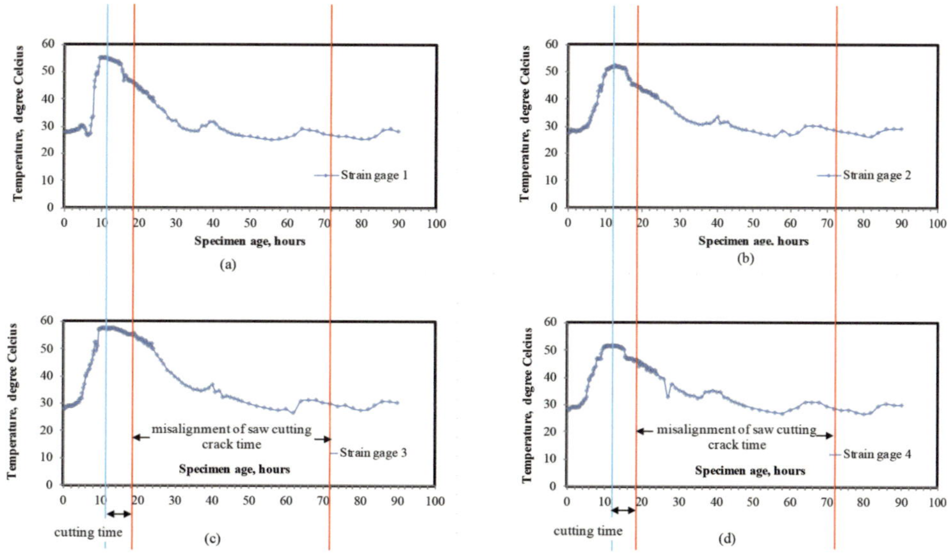

Figure 3.
Relationship between inner temperature and time of slab: (a) EVWSG 1, (b) EVWSG 2, (c) EVWSG 3, and (d) EVWSG.

According to the specification of the toll road, the cutting should be done in the time range of 12–18 hours that is when the rate of decline in the inner temperature is high (**Figure 3**). Cutting after this period results in misalignment cracks (**Figure 3**).

3. Concrete pavement

3.1 General over view

In this research, concrete pavement was done on the lean concrete plate. The aim of pouring on the plate is to sustain shear and tensile, which always appear under surface structure. Plate foundation can have different supported capacity [20].

The observation was done in the Bakauheni-Palembang road, Lampung Province, Indonesia, with 30,000 m length. The concrete pavement has four lanes (**Figure 4**). In this observation, the dowels have 32 mm diameter and 70 cm length, and the space between two dowels is 30 cm.

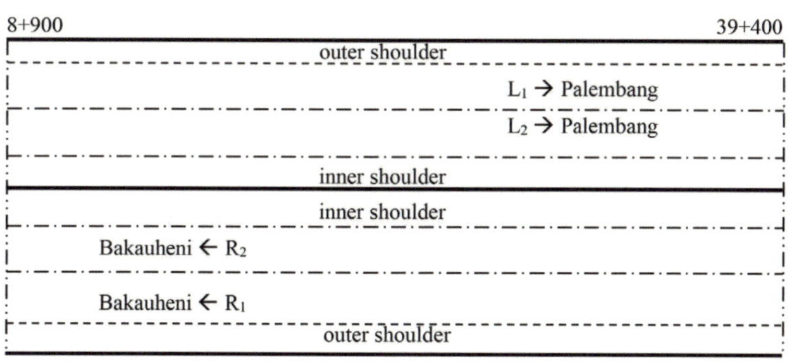

Figure 4.
Real concrete pavement scheme.

The lanes were divided into many segments. Each segment has 5 m length and 4.6 m wide; therefore, there are 600 segments for every lane or the total segments is 2400. Misalignment cracks appear in observation road (**Figure 5**).

A transverse crack in the middle of the span can cross the segment, but the misalignment crack does not cross to other segments. This happens because of different mechanisms of cracking. The types of defects observed that occur from 24,000 segments can be seen in **Table 1**.

Misalignment cracks are the second part of the three types of damage (**Table 1**). Eighty to ninety percent of the misalignment cracks occur on dowels that experience shifting or late cutting. The crack width is 1.5–2 cm with 30 cm depth or until the base of rigid pavement (**Figure 6**).

This damage is serious because it gets to the bottom of the concrete pavement. Understanding the mechanism of the occurrence of misalignment cracks is expected to trigger construction implementers to take preventative actions based on an understanding of the seriousness of the damage that will occur.

Figure 5.
Misalignment cracks: Bakauheni-Palembang.

Crack type	L segments	R segments	Total segments	% of total segments
Random	11	6	17	0.07
Misalignment	81	53	134	0.56
In the middle of the segment	98	94	192	0.8
Total			343	1.43

Table 1.
Percentage of crack type.

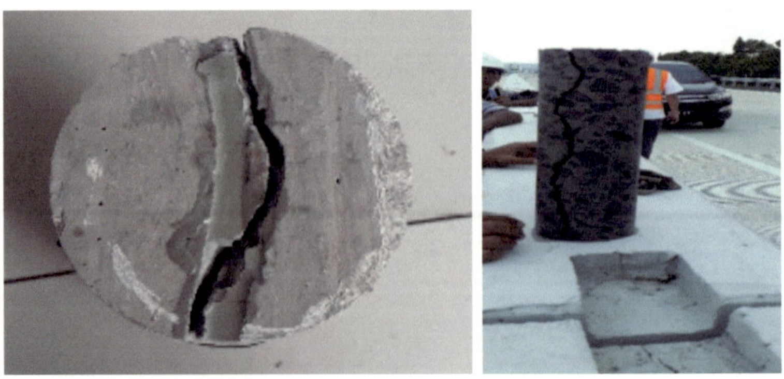

Figure 6.
Width and depth of misalignment crack.

3.2 Dowel

In general, concrete pavement is intended for heavy loads. To transfer the burden, dowels are required. Dowels made of plain reinforcement are installed between two segments (**Figure 7**).

The dowels were placed on the transversal bars, which formed as a chair (**Figure 8**). To maintain the chair and the dowel in the right position, some studs were placed by nailing it into lean concrete or LC (**Figure 8**). The dowel is mounted in the thick center of the rigid pavement (**Figure 9**). The lean concrete is 10 cm thick with compressive strength of 10 MPa (**Figure 9**). The rigid pavement has a bending strength of 45 MPa and is 30 cm thick. It was placed on the lean concrete covered with plastic membrant (**Figures 8** and **9**). A plastic membrant should be placed between the concrete/rigid pavement and lean concrete (**Figures 8** and **9**). The function of the plastic membrant was studied [7].

Pouring fresh concrete on the dowel should be done carefully, with little volume and slowly but continuously (**Figure 10**). This method is done to keep the dowel position still in right position. The dowel shift is caused by (1) the weakness of the stud installation so that it cannot hold the pouring of large volumes of fresh concrete at a location, or flattening and compaction by paver equipment (**Figure 10**) or (2) the weakness of binding dowel to basket.

When casting passes dowel line, the dowel position and its shift are recorded. Generally, the shifting dowel is between 10 and 20 cm from the correct axis (**Figure 11**).

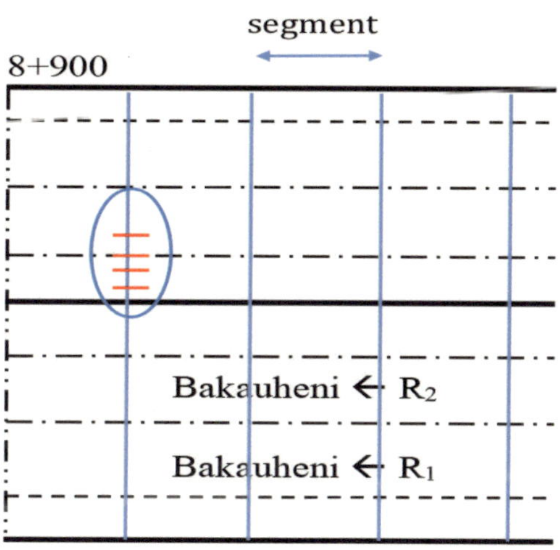

Figure 7.
Segment and dowel.

Figure 8.
Dowel baskets placed on an LC. Red line: Dowel. LC: Lean concrete.

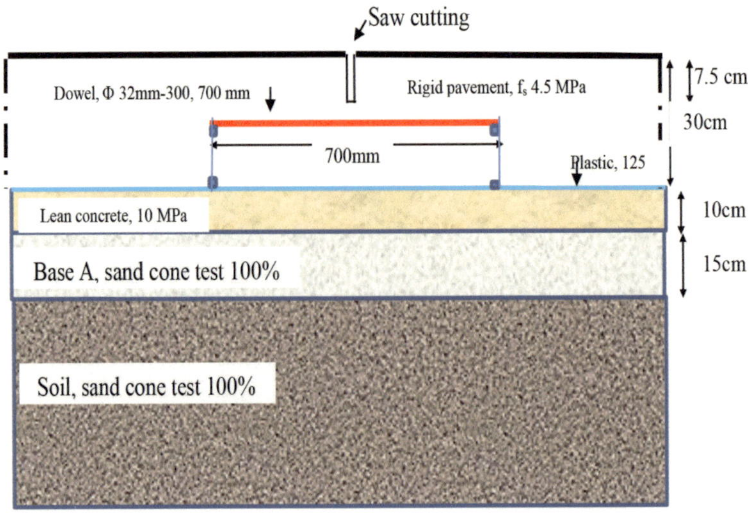

Figure 9.
Dowel position on cross-section.

Figure 10.
Pouring, flattening, and layering of the concrete/rigid pavement. Compaction process.

Figure 11.
Shifting dowel: (a) the top dowel holder shifts, and (b) the dowel holder shifts.

3.3 Saw cutting

Above the dowel line, a transverse cutting by saw is carried out, to direct the crack to occur at that place (**Figure 9**). Open free in concrete pavement by saw cutting means making the area weak. Cracking is planned to occur in the area (**Figure 12**).

If a crack exists below the saw cut, and an uncontrolled crack occurs nearby, then it is possible that dowels are misaligned and have caused locking of joint [18]. The depth of saw cutting varies, including D/2, D/3, and D/4. D is the thickness of rigid pavement. In this research, saw cutting was done above the as dowel line. The saw cutting depth d is D/4 (**Figure 13**). Cracks process and cracks area with various depths can be seen in **Figure 13a** and **b**.

Microcracking appears below the notch (**Figure 13a**). On cutting as deep as D/2, there is still a microcracking area that has not been covered (**Figure 13b$_1$**). This microcracking force will spread sideways and cause weakness in other areas. Whereas on cutting as deep as D/3, it can be said that undamaged areas are very thin (**Figure 13b$_2$**). On cutting as deep as D/4, the layer of area that protects against the distribution of microcracks is sufficient (**Figure 13b$_3$**). The distribution of concrete stresses under the saw cutting is as shown in **Figure 14**.

Concrete conditions when misalignment cracks are at the age of 18–72 are between plastic and brittle. Stress distribution at this time was assumed as shown in **Figure 14b** that is a transition between that shown in **Figure 14a** and **c**.

The stress distribution changes as the concrete maturity changes as shown in **Figure 15** [13]. Stress distribution as shown in **Figure 15** is distributed with cutting age 12 hours.

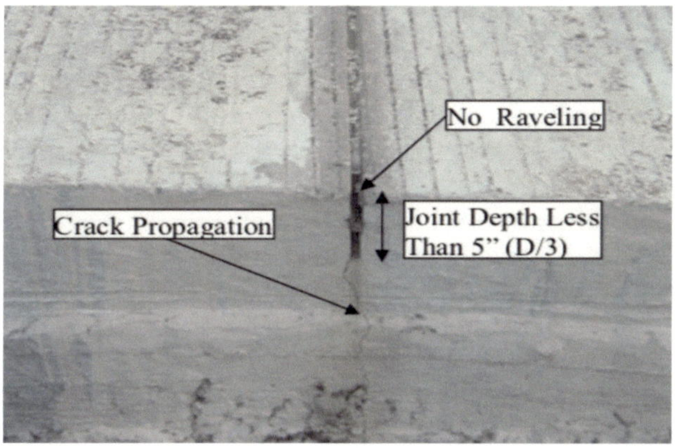

Figure 12.
Cracks occur at the planned place.

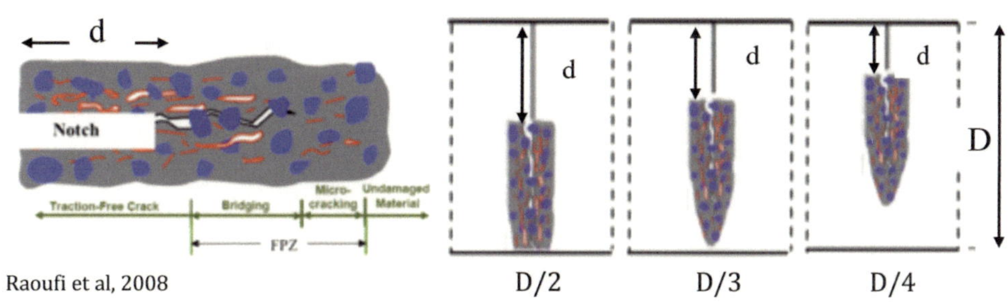

Figure 13.
Process and cracks area, (a) fracture process zone (FPZ), (b) FPZ in D/2, D/3, D/4 [13].

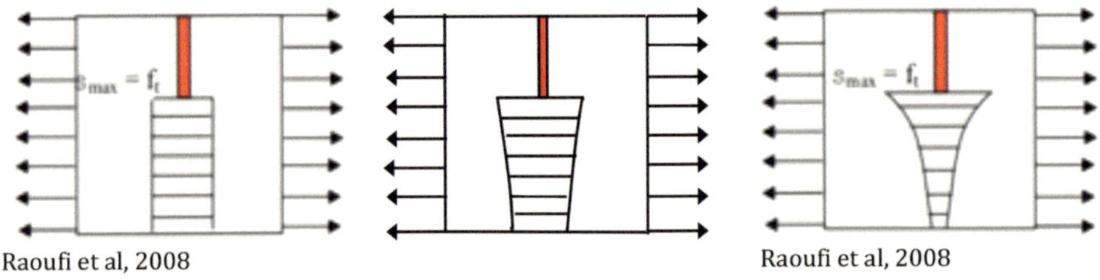

Figure 14.
Stress distribution: (a) plastic failure, (b) plastic-brittle failure, and (c) brittle failure [13].

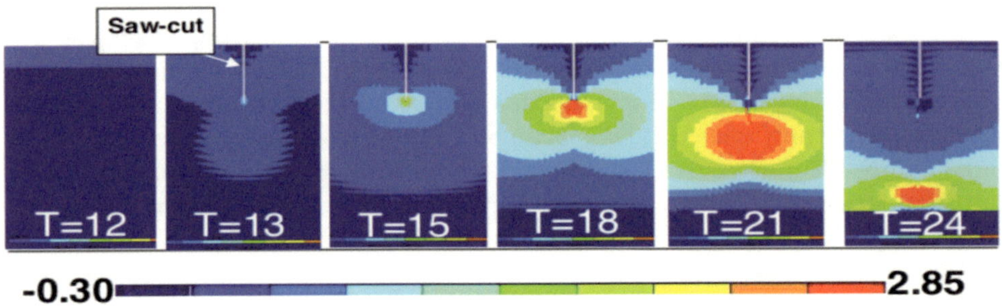

Figure 15.
Contour stresses at different times for rigid pavement with a D/3 cutting depth are carried out at 12 hours [13].

There is an optimum time to saw contraction joints in new concrete pavements, which is defined as the sawing window. It represents as short period after the placement of concrete within which concrete can be cut successfully before it cracks in an uncontrolled manner. If the sawing of the joints is started too early, then it may lead to raveling along the cut (**Figure 16a**). Concrete has not been able to withstand the pressure of the existing blade and microprestress. The jagged, rough edges are termed as raveling. Some raveling is acceptable if the widening of saw cut for filling joint sealant would remove the ravel edge. If the raveling is too severe, it will affect the appearance and ability to seal the joint. Saw cutting done at the beginning of the window time results in moderate raveling (**Figure 16b**) and at latter of cutting window time result good performance (**Figure 16c**). If sawing of joints is delayed beyond a certain period when significant concrete shrinkage occurs, then it may induce random cracks within the pavement.

Based on the results of the saw cutting, the appropriate time is at the end of the window time. Raoufi, 2008, [22] illustrates that excellent results will be obtained at fc'800–1000 psi (5.5–6.9 MPa). Saw cutting on rigid pavement according to the specification of this research should be done in the time range of 5–18 hours. At this age, the shrinkage begins to change into expansion (**Figure 2**) and the temperature inside the concrete has reached a maximum value (**Figure 3**). High temperatures cause the particles to move fast, resulting in collisions between molecules. This

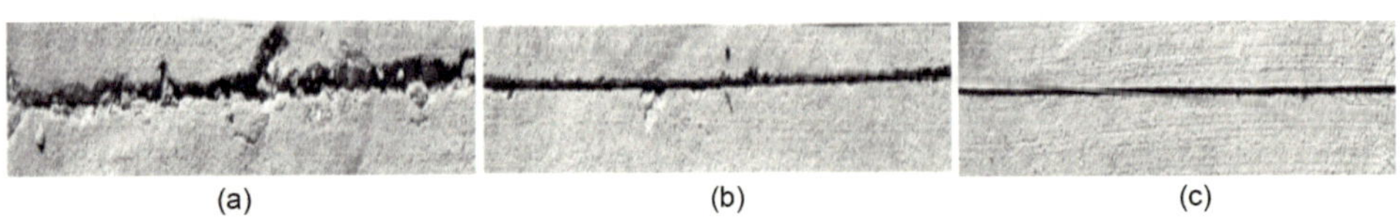

Figure 16.
Saw cutting result: (a) sawed too early (unacceptable), (b) sawed early in window (moderate), (c) sawed latter in window (acceptable) [21].

impact energy requires a place to release so there is no build-up. With saw cutting done in the window time, there is a space for expansion and shrinkage, so that cracking does not occur anywhere else. Notch from saw cutting is also a place for releasing energy.

Misalignment cracks as shown in **Figure 5** appear at the age of 18–72 hours. The crack follows the location of the shifted axle. Distance of the crack is about 50 cm from cutting line axle. Theoretical and practical studies have shown that the optimum joint spacing depends upon the slab thickness, concrete aggregate, subbase, and climate. Pavement with long transverse joint spacing may crack at locations other than the saw cuts due to tensile stresses from temperature curling. Most of the time, the spacing of transverse contraction joints in plain pavement is kept at 4.5–6.0 m. It is also important to check the transverse and longitudinal contraction joint spacing to see if it is within the limit as described in various codes and specifications. Saw cut transverse joint at proper location with respect to position of dowel bar assembly is a common construction practice. But some occasional mistakes may happen resulting in the misplaced saw cut. Proper location of saw cut dowel assembly not only improves the joint load transfer efficiency but also ensures better performance of the pavement throughout its life. Tolerance for sawing transfer joint, that is, the allowable translation of saw cut from the middle of the dowel assembly, depends upon the length of the dowel. It has been found that 15 cm of dowel embedment is all that is necessary for an effective load transfer under highway loadings. Thus, for a commonly used 50 cm long dowel bar, the available tolerance is 20 cm, that is, 10 cm either way from the center of the dowels.

4. Misalignment crack mechanism

To analyze the mechanism of crack misalignment, the influence of growth of hydration, decreasing pores, and microprestress needs to be understood.

4.1 Growth of product hydration and decreasing pores

The amounts of pores and the growth of product hydration can be seen in **Figure 17** [23]. t_c is the misalignment cutting time crack in the observed toll road.

The time frame for the appearance of misalignment cracks is expressed as t_c in **Figure 17**. During this time, the rate of hydration products' growth such as CSH, CH, C-A, F-H is very high; ettringite reaches its maximum and then decreases because part of it turns into monosulfate; and the number of pores decreases at a high rate.

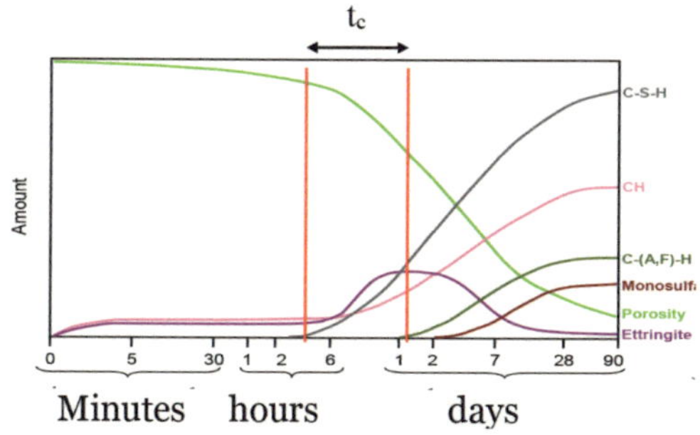

Figure 17.
Product hydration growth and decreasing pores.

4.2 Microprestress

Product hydration growth and decreasing pores cause stress in the concrete. The microstress is known as microprestress. Microprestress is the stress in micromeasurement, which occurs without applied load. Microstresses are due to the growth of hydration products, capillary pore pressure, disjoining pressure, and surface pressure [24]. Stress due to heat hydration, shear between fresh concrete and lean concrete, and shear between concrete and dowel are also classified as microprestress. Misalignment cracks in the concrete pavement occur at 18–72 hours (**Figure 17**). In that time span, the strain and temperature of the concrete decrease gradually (**Figures 2** and **3**). The growth rate of hydration products and the decrease of pore number rate in the span of time are very high (**Figure 17**). The consequence of this is that high stress occurs in the concrete matrix and in the pores (**Figure 18c**). This period is a critical time after the autogenous phase. Beside the microprestress, Buchta, 2015, [20] mentioned that the shear and tensile were concentrated on the base of the plate. This phenomenon will increase the microprestress. This is because the strain in the center of the plate is higher than in the surrounding area (**Figure 2**). The causes of microprestress will be explained below.

4.2.1 Product hydration growth pressure

The hydration process makes fresh concrete evolve into hard concrete (**Figure 18a** and **b**).

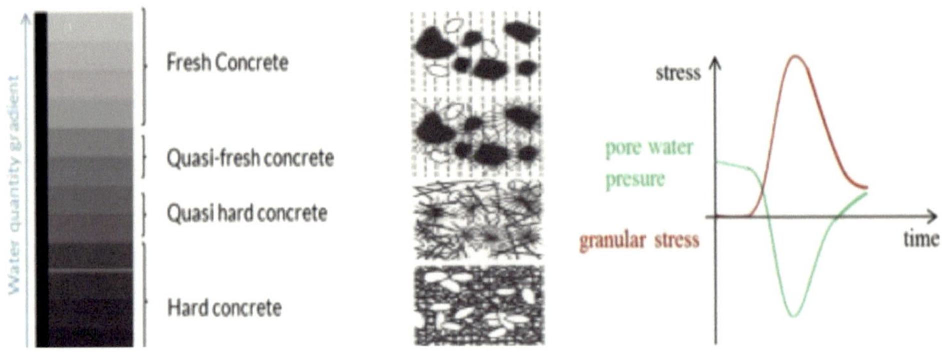

Figure 18.
Concrete evolution, (a) water quantity gradient, (b) microstructure evolution, (c) granular stress and pore water pressure [23, 25].

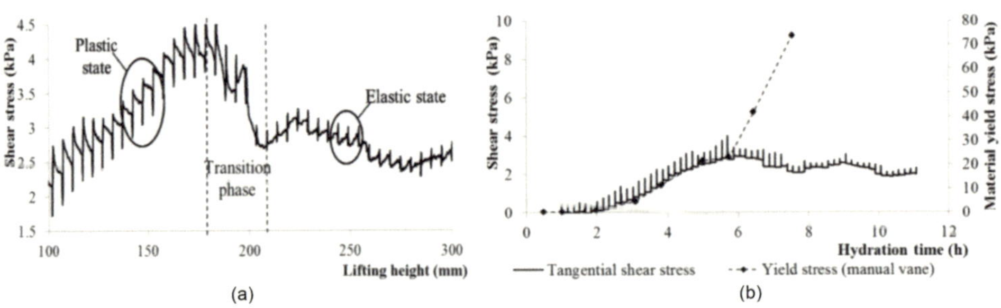

Figure 19.
Mortar behavior evolution: (a) behavior type, (b) material yielding [25].

Quasi hard concrete occurs after the age of 6 hours when the CSH growth rate increases (**Figures 17** and **18a**) and yield stress occurs (**Figure 19b**). The characteristic of quasi-brittle material is a strong nonstatistical size effect on both the structural strength and the rate of shrinkage and creep [25].

This evolution changes the deformation behavior of plastic behavior into a phase transition, and then changes to elastic state (**Figure 19a**). The behavior of microprestress in mortar or concrete also changes; at first microprestress is able to penetrate water at fresh and quasi-fresh concrete and then turn into bouncy due to hitting the quasi-hard and hardened concrete and also reinforces bar and dowel (**Figure 18a** and **b**). The bouncy power can break the weak bond and microcrack occurs. Concrete such as quasi-brittle structures fails at the macro-crack initiation, which can appear from the microcrack growth. Shear coefficient was also changed due to the increase of the concrete hardness and the power of microprestress. The change of shear stress can be seen in **Figure 19b**. The change from plastic state to transition phase is mentioned by Craipeau et al., 2018, as yield stress (**Figure 19b**) [25]. Concrete yield stress according to **Figure 19b** is about 22 kPa (0.022 Mpa) and occurs at the age of 6 hours. At the concrete age of 6 hours, C-S-H and ettringite begin to grow rapidly (**Figure 17**). If the dot line (**Figure 19b**) is continued, in the range of 18–72 hours, the shear stress is 320–1914.3 kPa (0.32–1.9 MPa).

Granular stress and pore water pressure show a similar value (**Figure 18c**). If it happens together, it will negate each other.

4.2.2 Evaporation

Evaporation occurs immediately after placement until the sample is tightly closed. Slowik et al., 2013, [26] found that although the rigid pavement was done by curing agent, evaporation still occurred (**Figure 20**).

In the first 30 minutes, there was a similarity of evaporation water between with and without curing, which is ±0.35 kg/m^2 (**Figure 20**). Evaporation without curing at 360 minutes (6 hours) is about 2.7 kg/m^2 or 2.7 10^{-6} MPa. Evaporation charts tend to be asymptotic after 360 hours (**Figure 20**); thus, evaporation at 3 days can be approached with 3 kg/m^2 or 3. 10^{-6} MPa without curing condition. This value was chosen because it shows the highest risk.

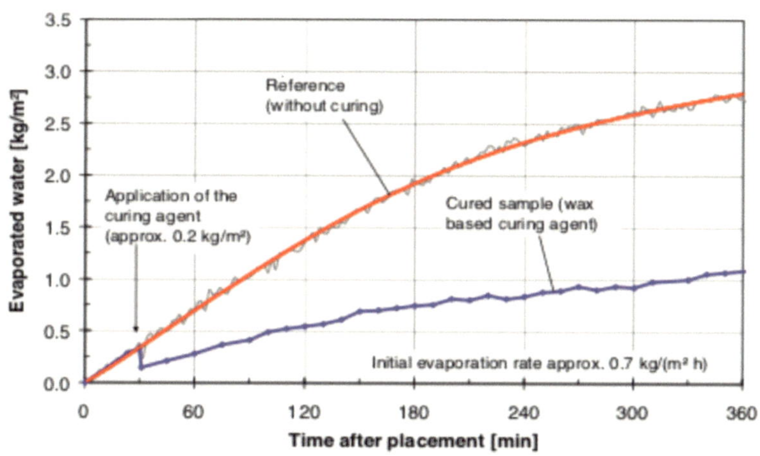

Figure 20.
Evaporation [9].

4.2.3 Capillary pressure

Water in the capillary pores causes tension pressure. Capillary pressure occurs quickly in concrete that contains more cement and less water (**Figure 21**).

By assuming that the capillary stress continues to run linearly, from **Figure 21b**, a capillary stress was obtained of approximately 163–828 kPa or 0.163–0.828 MPa in the age range of 18–72 hours. This figure is an estimate due to the influence of relative humidity.

The size and number of pores change due to the hydration process (**Figure 17**); therefore, capillary tension always occurs. The larger pores will dry and smaller pores will remain saturated. The surrounding relative humidity affects pore size (**Table 2**).

Because the surrounding relative humidity (RH) always changes, so does the capillary stress. Pore size determines the ability of water evaporation, menisci strength, and its properties. Pores with smaller sizes indicate that the smaller the evaporation ability, the smaller the evaporation stress [28]. Medium capillary pores (50–10 nm) and small gel pores (10–2.5 nm) have strong menisci [28]. The strong menisci mean strong capillary pores.

4.2.4 Disjoining pressure

Disjoining pressure is a pressure due to the attractive force between two surfaces, divided by the area of the surfaces; thus, disjoining pressure depends on the distance between the two surfaces. Correlation between inner water thickness and disjoining pressure was presented by Peng et al., 2015, as shown in **Figure 22** [30]. The thicker the inner water, the smaller the disjoining pressure. Maximum disjoining pressure is 50 bar (**Figure 22a**) or 5 MPa.

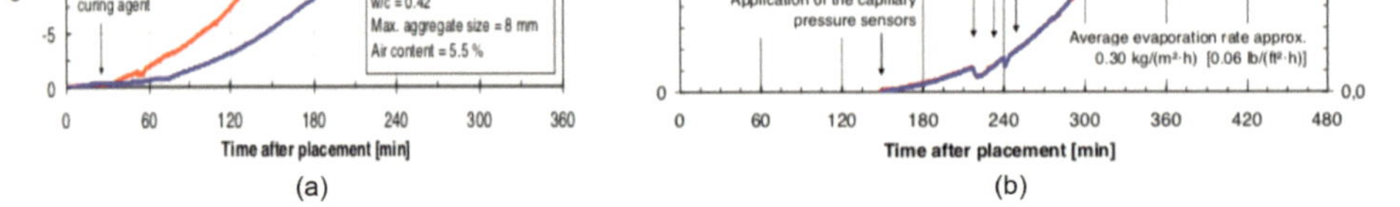

Figure 21.
Capillary pressure: (a) 420 kg/m², w/c 0.42; and (b) 335 kg/m², w/c 0.45 [27]. Source: Slowik et al. [22].

Relative humidity	85%	74%	54%	33%	<33%
Pore diameter (mm)	13	7.8	3.4	~1	<1
Origin of pores	Capillary pores	LD C-S-H (gel pores)		HD C-S-H	C-S-H structure
Source: Jenning et al., 2005 [18].					

Table 2.
Theoretical relationship between relative humidity and pore size.

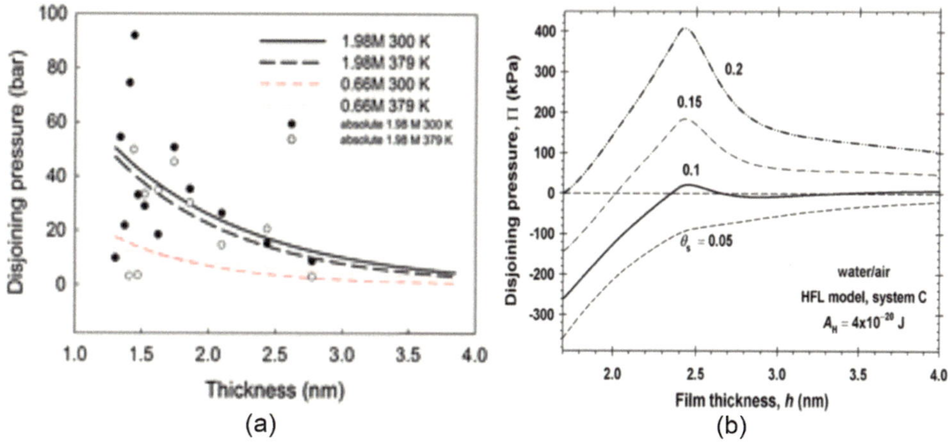

Figure 22.
Disjoining pressure: (a) disjoining pressure of water, and (b) total disjoining pressure for foam films and for the surfactant [4, 29].

4.2.5 Total microprestress

Total microprestress is taken from the description above. The values above are approach values not absolute values; it happens because there is evolution in the concrete that always occurs, and the influence of surrounding relative humidity makes microprestress to be always changing. The highest risk is anticipated by selecting extreme values from the description above (**Table 3**).

4.3 Dowel effect

There is friction force between dowel and concrete. The friction is not constant because the microstructure of concrete has evolved over time. Friction of steel in fresh concrete was observed [27]. Rabbat and Russel, 1985, have studied friction coefficient of steel on concrete or grout [10]. The effect of dowel on misalignment crack is divided into cutting at window time and cutting at out of window time with dowel in right position and shift position.

4.3.1 Cutting at window time

4.3.1.1 Dowel in right position

Placement of the correct dowel position and sawed in window cutting time result a crack occurring right in the saw cutting line (**Figure 12**). When saw

Type of microprestress cause	Maximum pressure, MPa
Product hydration growth	1.9
Evaporation	3.10^{-6}
Capillary pressure	0.83
Disjoining pressure	5
Total	7.730003

Table 3.
Maximum stresses from various causes of microprestress.

cutting is done, the frictional force between the dowel and concrete is balanced on both sides of the saw cutting line, penetrating the fracture that occurs and creeping. Microprestress penetrates cracks and then propagates under rigid pavement. Propagation of these forces in the area under the saw cutting increases the high stress that is collected at the lower end of the saw cutting line. This makes the saw cutting line the weakest area, so cracking occurs in the area (**Figure 23**).

4.3.1.2 Dowel in shift position

Dowel can shift from the right position (**Figure 11**) due to the pounding of fresh concrete pouring or because it is displaced by a paver (**Figure 10**). Naturally, there are shear force between dowel and concrete. Shifting dowel makes unbalanced shear force between both sides of the cutting line. Addition with microprestress and concentrated stress at the lower end of the saw cutting separates unevenly due to the dowel shift and increases unbalanced stress on both the sides. To be balanced, then when cutting arises the pulling force of restraint (**Figure 24a**). The consequence of this is that cracking does not occur in the cutting line but shifts at the starting point of the holding force.

The microprestress can reach 2.73 MPa as in item 4.2.5; if added dowel frictional force and stress concentrated at the lower end of the notch can exceed the strength of concrete that is 5.5–6.9 MPa (point 3.3). This can cause microcracks to arise in the concrete. The microcracks and high stress at one side of the cutting line lead to misalignment cracks (**Figure 24b**).

4.3.2 Cutting at out of window time

Saw cutting is done after the plastic period in which stress distribution is approached as shown **Figure 14b** and enters brittle condition. The brittle

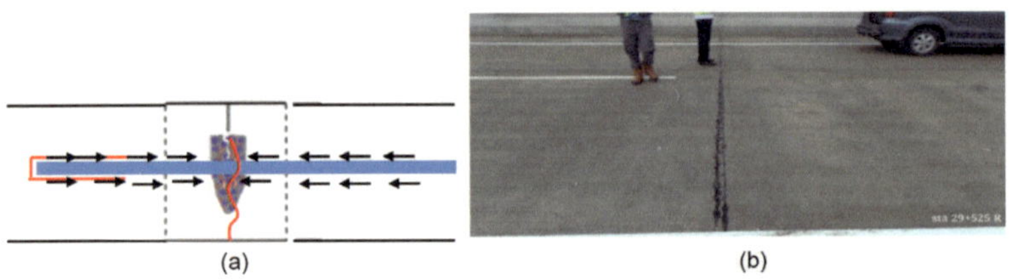

Figure 23.
Dowel in right position: (a) shear stress between dowel and concrete and right crack, (b) cross crack in right position.

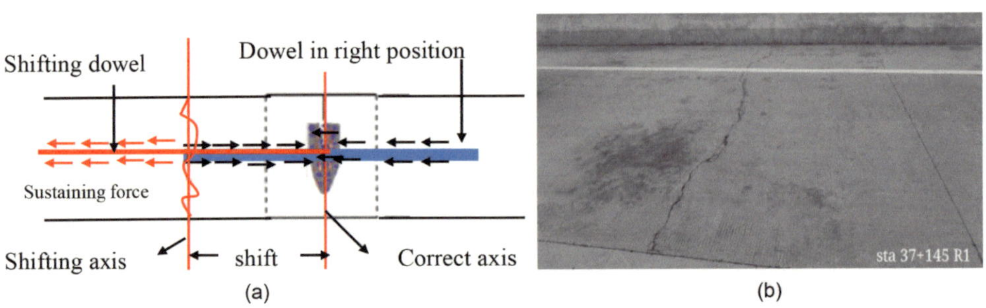

Figure 24.
Dowel in shift position: (a) shifting dowel and unbalanced stress, (b) misalignment crack.

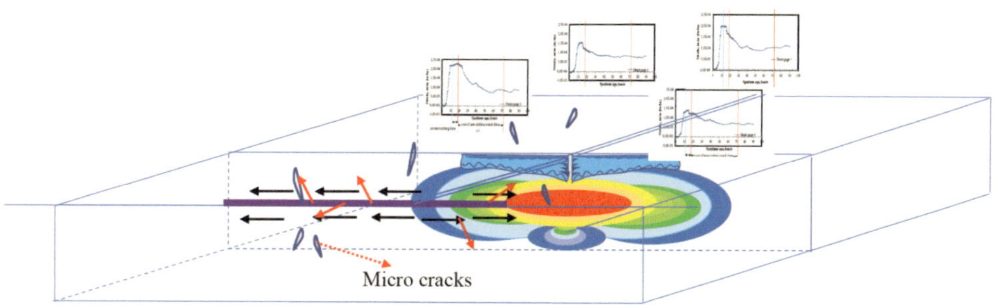

Figure 25.
Bouncy force of microprestress.

condition is marked by the end of ettringite growth. Based on this and **Figure 17**, it begins at around 18 hours of age. Distribution of stresses under the saw cutting is wider to both sides in brittle condition than in plastic condition (**Figure 14c** and **a**). Estimates of stress contours under brittle conditions are presented in **Figure 25**.

The center of the saw cutting line shows the smallest deformation (**Figure 2**), but surrounded by a larger strain (**Figure 25**). It happened at the age of less than 10 hours (**Figure 2**). The stress that causes this strain is withheld by the concrete if there is no room for expansion. Concrete bonds may still be able to withstand these stresses, but by cutting in brittle conditions, the restrained stress increases with a high stress that widens at the bottom end of the cutting. The consequence of this situation is a high-stress accumulation that occurs on both sides of the saw cutting line. As a result of this, accumulation on both sides of the saw cutting line is weak making it more vulnerable to crack.

4.3.2.1 Right position dowel

Concrete is a quasi-brittle material where the fracture process zone (FPZ) in the notch condition can be seen in **Figure 13**. In the FPZ area, microcracks are already present. According to the explanation as mentioned above, great stresses are created on the one side. The friction between the dowel and concrete, which is in a microcrack condition, does not run smoothly along the dowel because the frictional force partially penetrates the microcrack gap. Some microprestresses are bouncy that can crush the dowels and penetrate microcracks (**Figure 25**).

Because cutting is done too late, the large stresses are looking for a way to release the energy. The move of this energy triggers microcracks to be widened on the weak side and it creates a misalignment crack.

4.3.2.2 Shifting dowel

Asymmetrical conditions of stress occur on both sides due to the dowel shift. There have been weak points at locations outside the saw cutting line (**Figure 24**). As a result, transversal cracks can arise outside the saw cutting line. Condition as mentioned in 4.4.2.1. can also occur. Unbalanced condition between shear force between both the sides of cutting line increases by the spread of large stress at the end of cutting depth and unbalanced bouncy force on both the sides of the cutting line. The unbalanced force leads misalignment cracks to occur on the side of the shift direct dowel.

5. Resume

1. In the concrete pavement, there are microprestress forces, shear forces between dowel and concrete, microprestress resilient forces concerning dowels, shear forces, and tension forces at the base of the concrete pavement.

2. Microprestress is ±7.730003 MPa. The biggest contribution is disjoining pressure and the second is product hydration growth.

3. Disjoining pressure needs to be reduced by maintaining humidity in the concrete pavement at an early age because disjoining pressure will decrease if the water film thickness is large. Although maintaining high humidity makes evaporation also higher, the danger due to evaporation is smaller than the danger of disjoining pressure.

4. Expansion and shrinkage occur in concrete pavement alternately.

5. Free open space is needed to make space for the concrete pavement to expand and shrink and release stress that occurs by saw cutting.

6. Excellent result of cutting time will be obtained at highest inner temperature, which shows high product hydration growth.

7. Raoufi, 2008, illustrates that excellent results of cutting time will be obtained at f_c'800–1000 psi (5.5–6.9 MPa). It is difficult to know the concrete strength (f_c') at very early age. Compressive strength factor before 1 day has not been found. As a guide, cutting on toll roads in Indonesia is 12–18 hours.

8. If the strength of concrete is smaller than microprestress, microcracks occur in the concrete pavement.

9. Cutting in the window time raises the stress below the tip of the notch that is spread evenly until the base of concrete pavement.

    - Dowel in the right position: no misalignment cracks occur.

    - Dowel shifts: there are imbalance stresses resulting in cracks on the side where the dowel shifts.

10. Cutting late raises the stress below the notch tip and is greater than at the bottom of the concrete pavement. This stress will accumulate with microprestress, shear stress, tensile stress, and resilient stress.

    - Dowel in the right position: this tension will break a weak bond. If one of the bonds breaks because it is unable to withstand the accumulation of the stresses, the stresses will release its energy. Furthermore, the stresses move along the line where a large stress occurs on one side of the cutting line.

    - Dowel shift: an event equal to all of the above description coupled with an unbalanced dowel and concrete shear force creates a misalignment of cracks on one of the weak sides.

11. The implementation of concrete pavement requires the following:

    - Conformity to mix design regarding the amount and quality of material and environment.

    - Designers and managers need to have a good understanding of the consequences of neglect even if it is only one step. The implementation of concrete pavement requires high discipline at each stage. Negligence on one part, because of lack of understanding, and bad discipline or field condition can cause cracks. Any negligence during the construction process may lead to premature distresses in concrete pavement slabs.

6. Conclusion

Saw cutting done during window time causes the stress at the end of the notch to be spread evenly to the bottom of the concrete pavement. In addition to these stresses in concrete pavement, there are microprestress forces, dowels and concrete shear forces, microprestress resilient forces concerning dowels, shear forces, and tension forces at the base of the concrete pavement. Sufficient concrete strength is needed to withstand the accumulation of all these stresses, so saw cutting must be done at an adequate age but still in a condition between plastic and brittle.

The mechanism of the crack misalignment has been studied. The main cause of crack misalignment is that the cutting is done late, that is, after the window cutting time or in brittle conditions. This causes the stress not to be distributed evenly downward but large stresses occur at the lower end of the notch, which spread on both sides of the cutting line. This large stress will break the weak bond and the stress continues to travel along the cutting line because there is a large stress on the cutting line. The consequence of this is that there is a misalignment crack.

Dowel shift causes an imbalance of stress below the notch tip to increase. If the concrete is not able to withstand this imbalance, misalignment cracks arise even though the saw cutting is done at window time. Misalignment cracks tend to appear along the dowel shift.

Every mix concrete has special properties. We are suggested to test the inner temperature in plate sample with the dimension equal to one segment of real concrete pavement. The high inner temperature to be a guide to saw cutting time.

Acknowledgements

My gratitude goes to the Faculty of Engineering, University of Lampung, for supporting this research. I would like to give my highest appreciation to the Faculty of Engineering, University of Indonesia, for locating and maintaining test specimens.

Conflict of interest

"The authors declare no conflict of interest."

Author details

Chatarina Niken
Faculty of Engineering, Civil Engineering Department, University of Lampung, Bandar Lampung, Indonesia

*Address all correspondence to: chatarinaniken@yahoo.com

References

[1] Jagana R, Kumar CV. High-strength concrete. International Journal of Engineering Science & Research. 2017;**6**(2):394-407. DOI: 10.5281/zenodo.291853

[2] Jenning HM, Thomas JJ, Georgios C, Gevrenov JS, Ulm F-J. Nanostructure of CSH Gel in Cement Paste as a Function of Curing Conditions and Relative Humidity. In: Proceeding of the Concreep 7; September 12th. France: Ecole Centrale de Nantes; 2005. p. 7

[3] ACI Committee 360. ACI 360R-06. Design of Slab on Ground. American Concrete Institute; 2006. pp. 1-74. ASIN: B002R0QLVS

[4] Purnawan U. The influence of base course type on rigid pavement concrete strength. In: Proceeding of the 2nd Annual Applied Science and Engineering Conference (AASEC); 24 Agustus 2017; Bandung, Indonesia. IOP Conference Series: Materials Science and Engineering 288 (218). pp. 1-6. DOI: 10.1088/1757-899X/288/1/012120

[5] ACI Committee 209. ACI 209R-92 Prediction of Creep, Shrinkage, and Temperature Effects in Concrete Structure; 2002. pp. 1-47. ISBN: 9780870311222

[6] Al-Jumaili MA. Maintenance and repair of early cracking in airfield apron rigid pavement. Applied Research Journal. 2017;**3**(3):109-113. DOI: http://arjournal.org

[7] Niken C, Siswanto Y, Widodo, Tjahjono E. Cracking of open traffic rigid pavement. In: Proceeding of the SICEST Conference; 28 April 2016; Palembang, Indonesia; 2017. MATEC Web Conference 101, 05009. 2017. pp. 1-6

[8] Fairbain EMR, Azenha. Thermal cracking of massive concrete structure. In: Fairbain EMR, Azenha, editors. Hand Book of RILEM State of the Art Report of the RILEM Technical Committee 254 CMS. Springer. 2019. pp. 1-12. DOI: 10.1007/978-3-319-76617-1

[9] Well SA, Philips BM, Vandenbossche JM. Quantifying built-in construction gradients and early-age slab deformation caused by environmental loads in a jointed plain concrete pavement. International Journal of Pavement Engineering. 2006;**7**(4):275-289. DOI: 10.1080/10298430600798929

[10] Rabbat BG, Russell HG. Friction coefficient of steel on concrete or grout. Journal of Structural Engineering. 1985;**111**(3):505-515. DOI: 10.1061/(ASCE)0733-9445(1985)111:3(505)

[11] Sapountzakis EJ, Katsikadelis JT. Creep and shrinkage effect on the dynamic analysis of reinforced concrete slab-and-beam structures. In: Proceeding of the European Conference on Computational Mechanics; 31 August – 3 September 1999; Munich, Germany. 1999. pp. 1-17 https://www.researchgate.net/publication/280125775

[12] Mehta Y, Ali AW. Field cracking performance of airfield rigid pavements. Journal of Traffic and Transportation Engineering. 2017;**4**(4):380-387. DOI: 10.1016/j.jtte.2017.05.010

[13] Raoufi K, Radlinska A, Nantung T, Weiss J. Practical considerations for determining the time and depth of saw-cuts in concrete pavement. Transportation Research Record Journal of the Transportation Research Board. 2008;**2081**(1):110-117. DOI: 10.3141/2081-12

[14] Pradena M, Houben L. Influence of early-age concrete behavior on concrete pavement performance. Gradevinar. 2016;**69**(9):875-883. DOI: 10.14256/JCE.1931.2016

[15] Ioannides A. Stress prediction for cracking of jointed plain concrete pavements, 1925-2000: An overview. Journal of the Transportation Research Board. 2005;**1919**(1):47-53. DOI: 10.1177/0361198105191900106

[16] Chaddha S, Chauhan AS, Chawla B. A study on the rigid pavement construction, joint and crack formation. International Journal of Modern Trends in Engineering and Research. 2017;**4**(1):138-143. DOI: 10.21884/ijmter

[17] Chen DH, Won M. Field investigations of cracking on concrete pavement. Journal of Performance of Constructed Facilities. 2007;**21**(6):450-458. ISSN (print): 0887-3828, ISSN (online): 1943-5509. https://ascelibrary.org/doi/abs/10.1061/%28ASCE%290887-3828%282007%2921%3A6%28450%29

[18] Kumar B, and Mathur. Early Cracking of Concrete Pavement: Causes and Repair. 2019. Available from: https://www.nbmcw.com/tech-articles/roads-and-pavements/4988-early-cracking-of-concrete-pavement-causes-and-repairs.html

[19] ACI Committee 224. ACI 224R-0.1. Control Cracking of Concrete Structure. American Concrete Institute; 2008. pp. 1-47. ISBN: 978-0-87031-056-0

[20] Buchta V, Janulikova M, Fojtik. Experimental test of reinforced concrete foundation slab. Procedia Engineering. 2015;**114**:530-537. DOI: 10.1016/j.proeng.2015.08.102

[21] ACI Committee 325. ACI 325.12R-02. Guide for Design of Jointed Concrete Pavement for Streets and Local Roads. American Concrete Institute; 2002. pp. 1-32. ISBN: 9780870310768

[22] Slowik V, Schmidt M, Kässler D, Eiserbeck M. Capillary pressure monitoring in plastic concrete for controlling early age shrinkage cracking. Journal of The Transportation Research Board. 2014;**2441**(1):1-5. DOI: 10.3141/2441-01

[23] Kurtis K. Innovations in cement-based materials: Addressing sustainability in structural and infrastructure applications. MRS Bulletin. 2015;**40**(12):1102-1109. DOI: 10.1557/mrs.2015.279

[24] Bažant ZP, Fellow ASCE, Hauggaard AB, Baweja S, Ulm FJ. Microprestress- solidification theory for concrete creep. I: Aging and drying effects. Journal of Engineering Mechanics. 1997:1188-1194. ISSN (print): 0733-9399, ISSN (online): 1943-78. 1997

[25] Craipeau T, Lecompte T, Toussaint F, Perrot. Evolution of concrete/formwork interface in slip forming process. In: Wangler T, Flatt RJ, editors. Book Series of First RILEM International Conference on Concrete and Digital Fabrication- Digital Concrete. Springer International Publishing; 2018. p. 12. https://hal.archives-ouvertes.fr/hal-01881000. Print ISBN: 978-3-319-99518-2, Electronic ISBN: 978-3-319-99519-9

[26] Slowik V, Schmidt M, Friztzsch R. Capillary pressure in fresh cement-based materials and identification of air entry value. Cement and Concrete Composite Journal. 2008;**30**(7):557-565. DOI: 10.1016/j.cemconcomp.2008.03.002

[27] Vanhove Y, Djelal C, Magnin A. A device for studying fresh concrete friction. Cement, Concrete, Aggregate Journal. 2004;**26**(2):35-41. DOI: 10.1520/CCA11897

[28] Holzer L, Gasser P, Muench B. Quantification of capillary pores and hardly grains in cement paste using FIB-Nanotomography. In: Konsta-Gdoutos, Maria S, editors. Hand Book of Measuring, Monitoring and Modeling Concrete Properties. An

International Symposium Dedicated to Professor Surendra P. Shah, Northwestern University, USA. Springer. 2006. pp. 509-516. DOI: 10.1007/978-1-4020-5104-3_62

[29] Danov KD, Ivanov IB, Ananthapadmanabhan KP, Lips A. Disjoining pressure of thin films stabilized by nonionic surfactants. Advances in Colloid and Interface Science Journal. 2006;**128-130**:185-215. DOI: 10.1016/j.cis.2006.11.011

[30] Peng T, Firouzi M, Li Q, Peng K. Surface force at the nanoscale observation of non-monotonic surface tension and disjoining pressure. Physical Chemistry Chemical Physic Journal. 2015;**17**(32):20502-20507. DOI: 10.1039/c5cp03050a

Chapter 7

The Effect of Geometrical Factors on the Surface Pressure Distribution on a Human Phantom Model Following Shock Exposure: A Computational and Experimental Study

Maciej Skotak, Molly T. Townsend, Eren Alay and Namas Chandra

Abstract

Experimental data and finite element simulations of an anthropometric surrogate headform was used to evaluate the effect of specimen location and orientation on surface pressures following shock exposures of varying intensity. It was found that surface pressure distributions changed with local flow field disturbances, making it necessary to use data reduction strategies to facilitate comparisons between test locations, shock wave intensities and headform orientations. Non-dimensional parameters, termed amplification factors, were developed to permit direct comparisons of pressure waveform characteristics between incident shock waves differing in intensity, irrespective of headform location and orientation. This approach proved to be a sensitive metric, highlighting the flow field disturbances which exist in different locations and indicating how geometric factors strongly influence the flow field and surface pressure distribution.

Keywords: shock wave, shock tube, pressure measurements, human phantom, end effect, numerical simulations, impulse effect, dynamic pressure

1. Introduction

The shock tube is a convenient way to generate the shock waves in a controlled fashion, and it has been employed in various research areas for more than a century [1–6]. The design of a compressed gas driven shock tube includes three standard components: driver (breech) and driven sections with an optional end wave eliminator [7, 8]. The differences in the dimensions (volume of the breech, breech-to-test section diameter ratio, length of the driven section) and operation of the tube (type of driven gas, mechanism of driver gas release) have significant impact on the resulting pressure history measured inside of the tube [9]. The classical design of the shock tube employs the plastic or metal membranes, which is used to confine the driver gas from entering the driven test section. The driver section pressure is gradually increased until the point of mechanical failure of the membrane upon which the driver gas is entering the test section, pressurizing the ambient gas and resulting in the formation of the shock wave. Alternative designs employ membraneless drivers where the piston [10–13], or fast acting valve [14–16], are used eliminating the need for membrane replacement between consecutive tests. Both designs have been demonstrated to allow generation of shock waves with diverse magnitudes and characteristics. It is worth mentioning various instrumental factors, discussed in detail in our recent contribution [17], can affect the quality of recorded pressure waveforms and impact the interpretation of the experimental data.

In the biomedical field, research utilizing shock tubes to investigate mechanisms of blast TBI (bTBI) was invigorated only 20 years ago [18, 19]. The primary goal in this area of research is to replicate conditions associated with field explosions, particularly the primary blast injuries caused exclusively by the interaction of a shock wave with the brain are of interest [20]. Simulation of explosive blast implies that a shock wave closely resembling the Friedlander waveform should be produced and it has become a standard in contemporary bTBI models [7, 21]. Two experimental parameters of paramount importance are the specimen restraint and the location of the test section, where the specimen is exposed to a shock wave. For the inanimate specimen the method of restraint is usually not an issue, however, human phantom models are frequently mounted on a biofidelic neck, e.g., Hybrid III, and when subjected to a shock wave loading rapid acceleration can result in the specimen displacement affecting the pressure loading on the surface. When the animal models are used the head restraint becomes extremely important, particularly for rodents with relatively small body dimensions and low weight. If the proper head restraint measures are not included in the experimental design, it might result in the development of erroneous injury modalities. It's been demonstrated that head acceleration might lead to the development of tertiary blast injuries, which have different injury characteristics than those resulting from shock wave loading [22, 23]. The importance of the test location in the shock tube was a subject of the experimental evaluation in the past by our group [24–26]. These results illustrate significant differences between testing the specimen inside of the shock tube, i.e., at the distance from the exit enough to eliminate influence of any end-effects, versus at the end and outside. Testing outside is undoubtedly more convenient, but carries a number of unwanted drawbacks: (1) the shock wave profile is eroded and typically only short duration waveforms are achievable, (2) there is large dynamic pressure component which might contribute to a variety of errors [27], (3) the loading of the specimen strongly depends on the location with respect to the shock tube exit and diameter due to highly heterogeneous conditions [28, 29]. Testing inside provides much higher level of control over the shock wave profile with the dynamic components resembling these encountered in the field explosions.

Numerical simulations are invaluable tools for mechanistic investigation of short lived phenomena like shock wave interaction with complex biological structures. The numerical models in bTBI research area provide insight into: (1) the transmission and propagation of the blast waves in the brain [8, 30, 31], and (2) mitigation of blast effects by helmets [32–40] or other PPE designed to safeguard the craniofacial area [36]. However, the accuracy of the numerical simulations relies on the validation using high-quality experimental data, e.g., the pressure measured on the surface of the helmet or phantom [8, 35, 39], or intracranial pressure [31, 40]. A related branch of experimental work which is yet to be explored to its full potential for numerical model validation is the use of post-mortem human specimens (PMHSs)

[37, 41, 42], instrumented with surface and intracranial pressure sensors. Existing studies in this area are rare and hindered by the experimental difficulties, not to mention complete impracticalities for the evaluation of PPE performance. It only leaves the use of anthropometric phantoms as the only alternative, which, made of non-biological materials, replicate the geometry to a high degree [8, 31, 39, 43, 44].

In this contribution we performed a comprehensive experimental characterization of the human phantom model instrumented with 10 pressure sensors to measure the response to a shock wave loading. The specimen was tested using large cross-sectional area shock tube using three locations with characterized by divergent flow characteristics. The loading of the specimen was administered via a single shock wave with three nominal intensities (70, 140 and 210 kPa), and surface pressure was probed in three headform orientations with respect to the incident shock wave, i.e., 0, 90 and 180°.

2. Methods

2.1 The shock tube

The 7 m long square (0.71 × 0.71 m) cross section shock tube was used in all experiments. This device was previously characterized in detail [24, 26]. The driver gas was compressed helium (ultra-high purity, 99.99%, Airgas, Oakland, NJ), which was allowed to flow into the breech separated from the driven section of the shock tube with membranes made of Mylar (Grafix, Cleveland, OH). Upon the rupture of membranes, the driver gas enters the driven section and compresses the ambient air, which in turn generates a shock wave. Three discrete Friedlander waveform shock waves with nominal intensities of approximately 70, 140 and 210 kPa peak overpressure in the test section (T5 sensor, **Figure 1**) were used. All tests were performed at ambient conditions.

2.2 Pressure measurement, headform preparation, and instrumentation

The temporal evolution of the incident shock wave waveforms was recorded using seven high frequency response pressure sensors model 134A24 (PCB Piezotronics, Inc., Depew, NY, USA), distributed along the shock tube (**Figure 1**). The pencil probe model ICP® 137B24B (PCB Piezotronics Inc., Depew, NY, USA) was used to measure the incident pressure on the outside (PP location, **Figure 1**).

The phantom headform [45], was instrumented with 10 PCB Piezotronics model 102B06 pressure sensors as illustrated in **Figure 2A**. Five medial sensors are located along midline anterior–posterior (H1–H5), and five circumferential sensors: two on the right parietal side (H6 and H7), two in eye sockets (H8 and H9, **Figure 2A**) and one on the left parietal side (H10). These sensors were mounted flush to the surface using tapped holes. The headform was mounted on the Hybrid III neck (Humanetics, Plymouth, MI) [46], in a rigid configuration to eliminate the motion of the headform during shock wave impact. The FOCUS headform-Hybrid III neck the assembly was attached to the adapter plate and bolted to the bottom of the shock tube in the test section in three different locations (**Figure 1**).

A custom LabView program was used to record the pressure waveforms. The data acquisition system is based on PXIe-1082 PCI Express chassis and PXI-6133 S Series multifunction DAQ modules (National Instruments, Austin, TX, USA). The signal of pressure sensors was filtered using 8-channel signal conditioners model 483C05 (PCB Piezotronics Inc., Depew, NY, USA). The pressure waveforms were recorded at 1.0 MHz sampling frequency with an acquisition time of 50 ms.

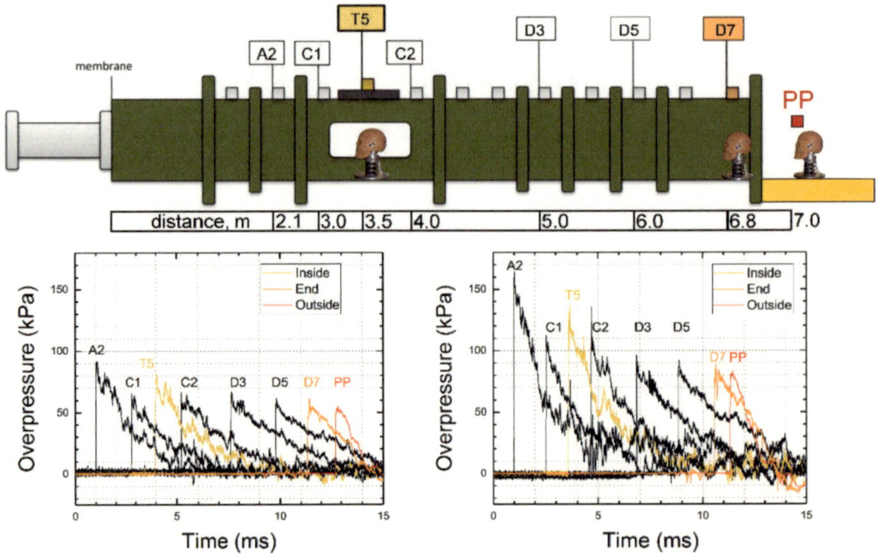

Figure 1.
Evolution of the shock wave profile traveling in the 28-inch square cross section shock tube. Top panel: schematics of the pressure sensor distribution along the shock tube. The headform is placed in three locations marked as: T5 (inside), D7 (end) and PP (outside). Bottom panels: Pressure profiles of the incident shock wave with nominal BOP of 70 kPa (left) and 140 kPa (right) recorded by sensors distributed along the shock tube as schematically depicted by the top diagram. The shock wave profiles recorded at T5, D7 and PP were used for experimental data normalization. This image was taken from reference [26] (distributed under creative commons attribution (CC BY) license).

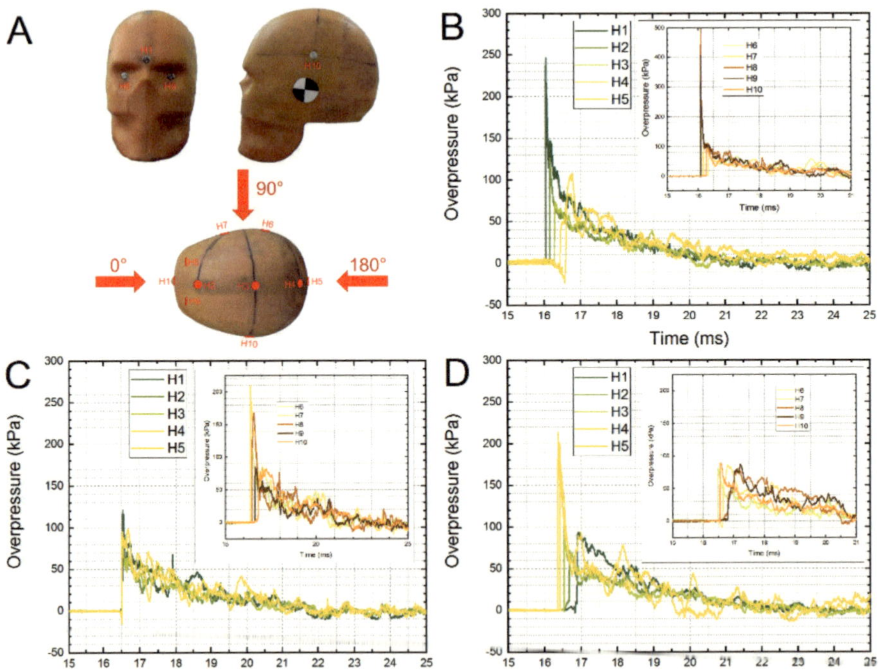

Figure 2.
The schematic representation of the pressure sensor distribution on the headform (A). The direction of the shock wave propagation in experiments where the headform was rotated in 90° intervals is illustrated. The experimental pressure waveforms collected at 70 kPa nominal shock wave intensity by the surface sensors to three headform orientations: (B) 0°, (C) 90°, and (D) 180° with respect to the direction of the shock wave propagation.

2.3 Experimental design, data reduction and statistical analysis

The study was designed as three-factor experimental design, 3 × 3 × 3. The three experimental variables investigated are: (1) the headform location (three levels: inside, end and outside, **Figure 1**), (2) the headform orientation (three levels: 0, 90, and 180°), and (3) shock wave intensity (three levels: 70, 140 and 210 kPa).

All pressure waveforms were processed and quantified in Origin 2018 software (OriginLab Corp., Northampton, MA). All data are presented as mean and standard deviation ($n = 4$).

The data normalization and reduction were performed as follows. The four pressure waveform characteristics were tabulated. These include (1) the peak overpressure, the increase in pressure observed at arrival of the shock front, (2) the rise time, or the time required for the pressure to increase from 10 to 90% of the peak overpressure, (3) the positive phase duration, or the time required to return to ambient pressure, and (4) the impulse, or the area under the pressure–time curve during the positive phase duration. Each pressure waveform characteristic value was divided by the incident shock wave equivalent (Eq. (1)):

$$x_{\mathrm{non}} = \frac{x_{\mathrm{p}}}{x_{\mathrm{i}}} \tag{1}$$

where, x_{p}—peak overpressure, rise time, duration or impulse of the resulting pressure waveform on the headform, x_{i}—peak overpressure, rise time, duration or impulse of the incident shock wave, at the headform location (T5, D7 or PP, **Figure 1**). As a result, a set of normalized non-dimensional peak overpressure, rise time, duration and impulse values is generated for each test location, headform orientation and shock wave intensity.

2.4 Flow-field simulations

Finite element models were used to identify the influence the three experimental variables on the flow-field around the specimen. The flow-field around the headform was simulated using a coupled Eulerian-Lagrangian approach to fluid-structure interaction. This solution technique is ideal for the simulation of large deformation finite element analyses and has been used extensively in the simulation of shock-structure interaction physics [8, 47, 48]. The shock wave is modeled in a Eulerian mesh of air and interacts with the headform, modeled using a Lagrangian mesh. The interaction of the two domains results in a solution which depicts the air flow around the headform.

The headform Lagrangian model was generated from a 3D geometrical model generated from the FOCUS headform and the Hybrid III neck using Autodesk Recap Pro 2018 (Autodesk, Inc.). The three-dimensional geometry was meshed using linear tetrahedrons (Simpleware, Synopsys) at a converged mesh density with an average minimum edge length of 8.73 mm (**Figure 3A**). The headform was assumed to be a linear elastic material with a density of 2700 kg/m^3, an elastic modulus of 6.89 GPa, and a Poisson's ratio of 0.33.

A pre-existing, validated model of a shock tube was enhanced for use as the Eulerian domain [49]. This model used a biased linear hexahedral mesh which converged at a minimum edge length of 14 mm at the region of interest. Two shock tube meshes were used, designed to best model the inside specimen placement and to model the exit and outside specimen placements. The model used to simulate the inside of the shock tube was 5.9 m in length and over 580 thousand elements (**Figure 3B**). The model used to simulate the end and outside specimen placements was 2.9 m in length and over 2.24 million elements (**Figure 3C**). The Eulerian mesh

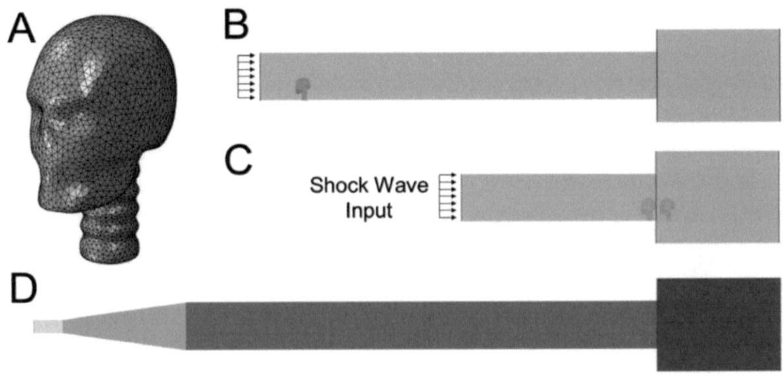

Figure 3.
The finite element models used in the calculation of the flow-fields around the headform. (A) The Lagrangian domain of the FOCUS headform was meshed using linear tetrahedral elements. The Eulerian domains were of two different lengths: (B) a 5.9 m length section downstream of the A2 sensor ***(Figure 1)*** *to model the inside specimen location and (C) a 2.9 m length section downstream of the D5 sensor* ***(Figure 1)*** *to model the end and outside specimen locations. (D) The full shock tube size is shown, depicting the breech, expansion cone, driven region, and catch tank.*

was assumed to be filled with air, approximated using the ideal gas equation of state at 296 K, a density of 1.2 kg/m^3, and a specific heat ratio of 1.4.

Twenty-seven simulated configurations were conducted, mirroring the experimental configurations of three specimen placement locations (inside, at the exit of, and outside the shock tube), three specimen orientations (0, 90, and 180°), and three blast overpressures (70, 140, and 210 kPa). The shock was simulated as a planar pressure wave, applied to a surface upstream from the headform. For each configuration, pressure measurements taken at the sensor coincident with the loading surface were averaged for the four repeated exposures to create an input incident pressure pulse. The pressure pulse for a 210 kPa exposure with the headform located at the exit was used as the pressure pulse for the untested configuration of 210 kPa with the headform outside of the shock tube. All nodes coincident to the walls of the Eulerian domain and the base of the Lagrangian headform were constrained against all translational and rotational degrees of freedom. The enhanced immersed boundary method allowed for the Lagrangian mesh to occupy void regions within the Eulerian mesh, enabling the computation of the interfacial surface. Interaction between the two domains was defined as hard general contact in the direction normal to the interacting surfaces and a frictionless contact in the tangential direction.

All simulations were conducted in Abaqus 6.13-4 on two Intel Xeon 2.10 GHz processors (Dassault Systèmes). For each configuration, the pressure in each element along the vertical longitudinal plane was mapped to plot the blast overpressure (BOP), impulse, and rise time amplification factors (MATLAB R2019a, Mathworks). Element-wise values for the peak BOP were defined as the maximum simulated pressures, the impulses were calculated as the area under the pressure-time curves, and the rise times were calculated as the time required for the signal to increase from 10 to 90% of the peak value. Element-wise values were normalized with values of the incident waveform at that location, resulting in unitless amplification factors.

3. Results

3.1 Experimental surface pressures on the headform

Representative pressure profiles recorded by the surface sensors mounted on the headform are presented in **Figure 2B-D**. These data were recorded at a nominal shock wave intensity of 70 kPa, and general trends in pressure waveforms distribution are similar for the other two incident shock wave intensities (140 and 210 kPa). The effect of the headform rotation on the surface loading is illustrated in **Figure 2** for the inside test location. In general, on the face exposed to the shock wave the recorded peak overpressures are the highest: 245 kPa (H1 sensor) for the 0° orientation (**Figure 2B**, H8 and H9 sensors are a special case, considering the sensors are located in the concave "eye sockets" which results in the pressure entrapment, and extremely high peak overpressure values), 220 kPa (H7 sensor) for 90° orientation (**Figure 2C**, inset), and 225 kPa (H5 sensor) for 180° orientation (**Figure 2D**). It is obvious that the same trends are observed for the other two test locations and headform orientations (**Figure 4A–C**). It appears that the rise time is a sensitive metrics for the headform loading (**Figure 4D–F**). In general, the rise time is shortened on the front face exposed to the shock wave (H1 at 0° orientation, **Figure 4D**, H6 and H7 at 90° orientation, **Figure 4E**, and H5 at 180° orientation, **Figure 4F**) or it is extended on the converse side of the headform (H4 and H5 at 0° orientation in **Figure 4D**, H10 at 90° orientation in Figure 4E, and H1 and H2 at 180° orientation in **Figure 4F**), compared to the rise time of the incident shock wave. This effect is seen regardless of the headform test location. For the duration and impulse, a different kind of relationship is noticeable. It is expected considering both peak overpressure and the rise time are describing the behavior of the front face of the waveform, while the duration and the impulse are related to the properties of the entire waveform. In general, the duration and impulse values decrease with distance from the breech, corresponding with the test location in the order: inside > end > outside. This is associated with the erosion of the incident wave tail by the end effect. The shock wave exiting the shock tube creates a region of underpressure

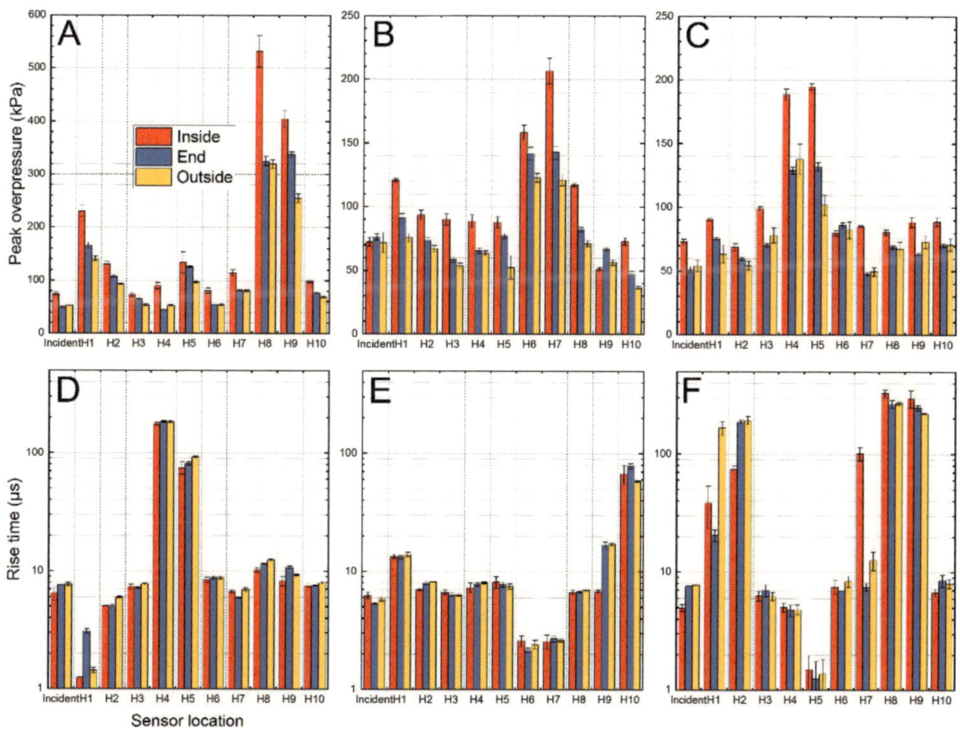

Figure 4.
The results of the pressure waveforms characteristics quantification. The peak overpressure and the rise time for the incident and surface mount sensors collected at three different test locations (inside, end and outside) are presented (n = 4). The headform orientation in these experiments was 0° (A, C), 90° (B, E), and 180° (C, F).

which travels back into the shock tube [7], and unconfined conditions on the outside allow for free expansion of the previously constrained shock front, resulting in a conversion of the static pressure to 'jet wind', resulting in a shorter durations and lower impulse values at the end and outside locations (**Figure 5**).

3.2 Experimental data normalization

It is thus obvious a further data reduction is necessary in order to compare data collected at three different intensities and three different headform locations. The simplest and most natural approach is to take the characteristic parameters of the incident shock wave waveform (input) recorded by a sensor mounted at specific test location, i.e., T5 for inside, D7 for end and PP for outside location, and compare them with waveforms recorded by pressure sensors on the headform (output). The resulting dimensionless parameters (calculated using Eq. (1)) are a measure of the disturbance caused by introduction of headform into the flow field of the shock wave traveling in the shock tube. The values other than 1 indicate divergence from the waveform characteristics at specific test location in or outside of the shock tube and on the specific location on the headform, compared to the incident shock wave. These can be attributed to geometric factors, changes in shock wave characteristics and presence of additional high velocity flows which are below the detection levels of small cross section sensors sparsely distributed on the surface of the headform. However, if the shock wave properties were only gradually evolving while traveling in the shock tube, it is reasonable to expect similar distribution of non-dimensional parameters as a function of their physical location on the headform which becomes the only defining parameter of the system.

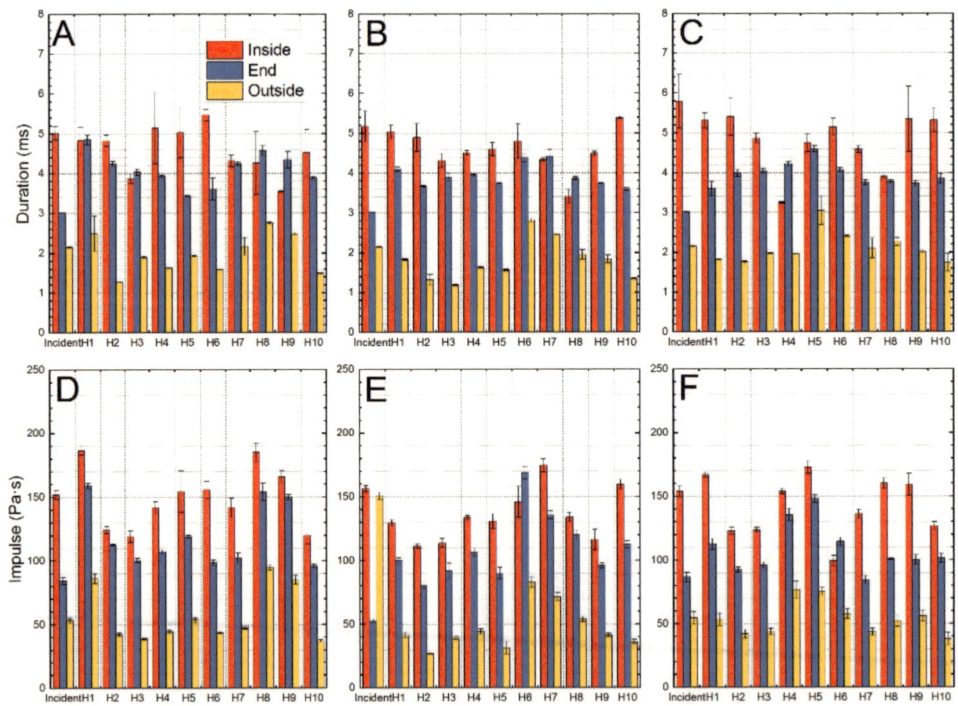

Figure 5.
The results of the pressure waveforms characteristics quantification. The peak duration and the impulse for the incident and surface mount sensors collected at three different test locations (inside, end and outside) are presented (n = 4)*. The headform orientation in these experiments was 0° (A, C), 90° (B, E), and 180° (C, F).*

The non-dimensional parameters (or amplification factors) allow for direct comparison of pressure waveform characteristic parameters generated by a range of incident shock waves differing in intensity. Using this concept in mind we performed further data reduction for peak overpressures, rise time, impulse and durations in all datasets. The representative bar plot for the amplification factors for the peak overpressure, rise time, duration and impulse for the headform in the 90° orientation and exposed to a shock wave with three nominal intensities (70, 140 and 210 kPa) in three different locations (inside, end and outside) is presented in **Figure 6**. This figure illustrates that indeed the normalization is an effective strategy to compare the data obtained from a variety of exposure conditions. The normalized peak overpressure and rise time follow well defined trends, and the largest divergence is observed for the normalized duration and impulse at the end test location (**Figure 6C** and **D**). It is expected considering duration and impulse values for the end and outside locations vary more significantly than these reported by the headform sensors in the inside location (**Figure 5**).

We previously reported similar trends for the headform tested in the 0° orientation (see Figure 7 in Ref. [26]), and this work expands upon these results by incorporation of two additional orientations (90 and 180°). The data are presented as a function of the sensor distance of the 2D projection of the headform (for details refer to Figure 7B in Ref. [26]). The best results were noted for amplification factors of peak overpressure recorded on the headform at 0° orientation (**Figures 7, 8** and **Table 1**): the data collected in all three test locations has narrow distribution and follows well defined trend. Front sensor H1 has increased values in the range of 2.4–3.6, which gradually decrease along the headform reaching minimum values

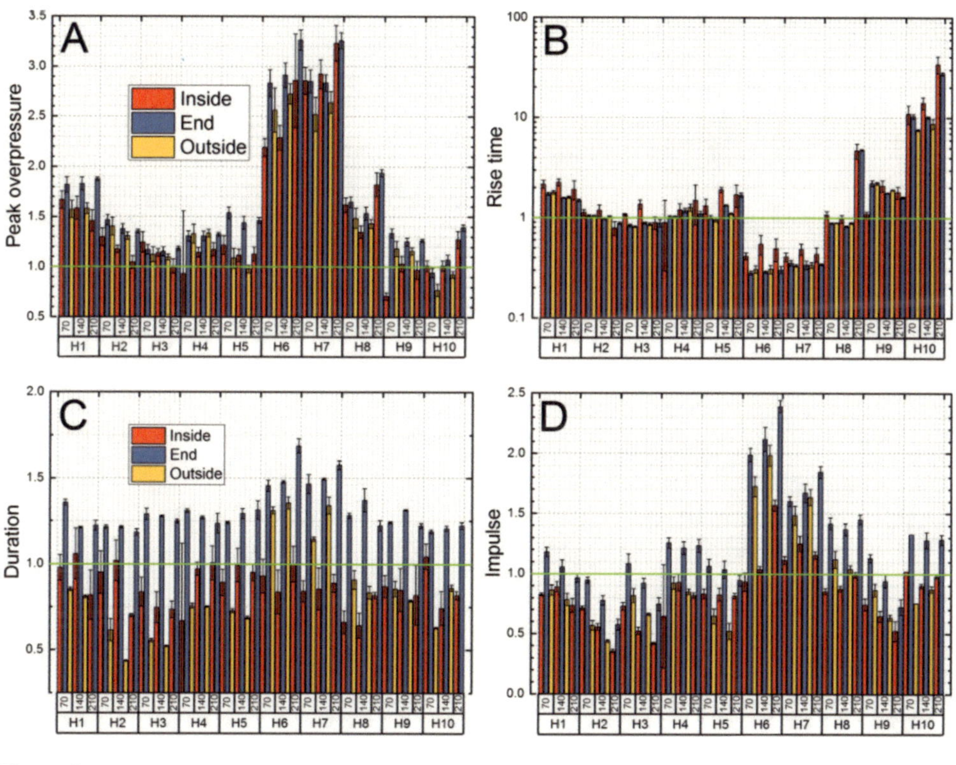

Figure 6.
The results of the pressure waveform characteristics quantification for headform mounted sensors. The normalized peak overpressure (A), rise time (B), duration (C) and impulse (D) are presented (n = 4) for experiments where the headform was mounted at 90° orientation. The normalization allows comparison of the data quantified for three nominal shock wave intensities: 70, 140 and 210 kPa.

of about 1 for the H4 and H6 sensors on the back of the headform, and reaching a value of 2 for the H5 sensor at the very end of the headform. This increase is purely due to combined effect of the shock wave wrapping around the headform and its two streams joining together at the back. This is accompanied by increased values for the rise time for H5 sensor by a factor of 10 compared to incident shock wave, which is markedly increased compared to all other sensors, where the rise time amplification factor never exceeds the value of 2 (**Figure 7**). Amplification factors for sensors mounted in the eye sockets are extremely high (in the range of 5–8),

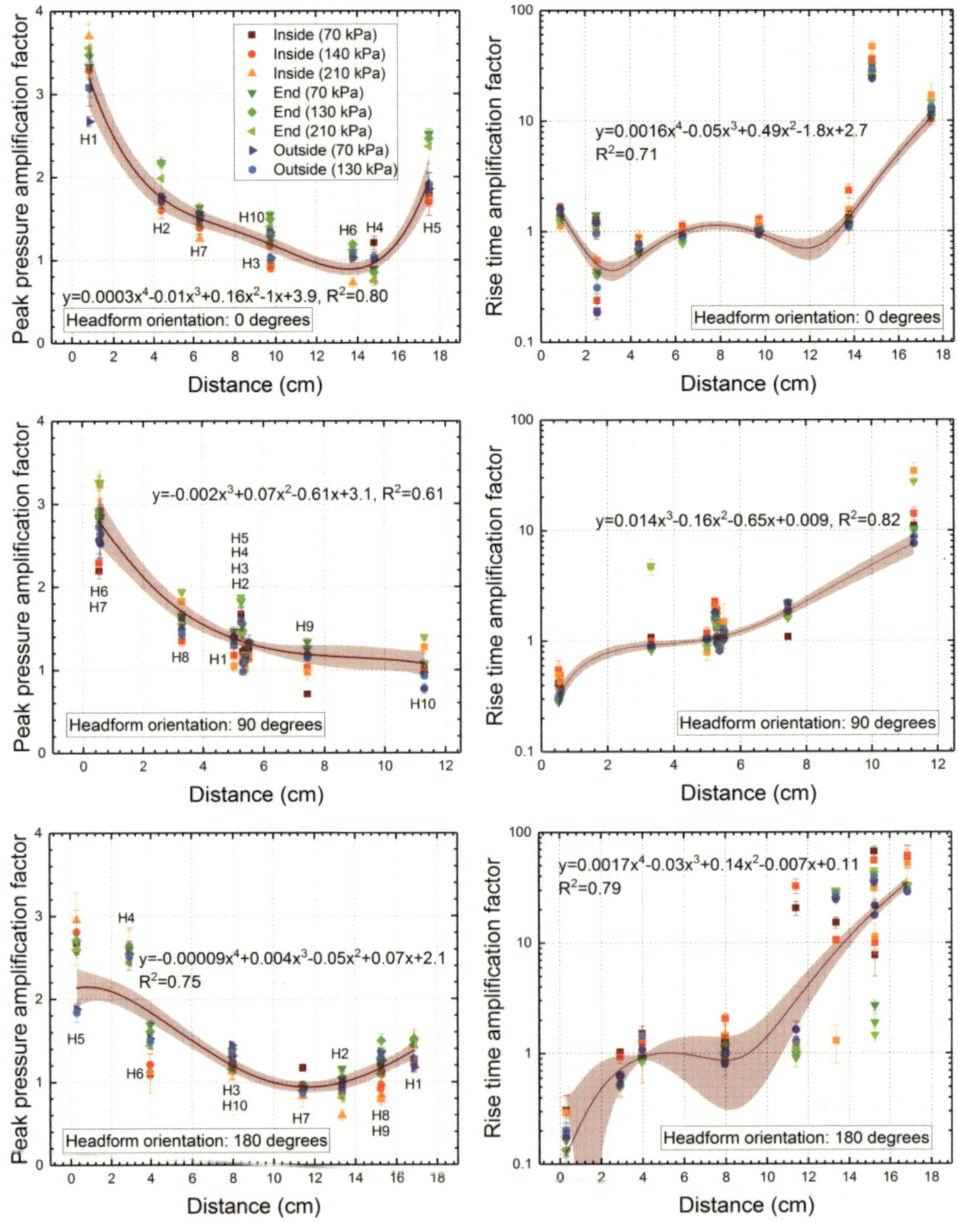

Figure 7.

Amplification factors for peak overpressure and rise time as a function of the headform sensor distance. Polynomial fit performed on normalized data collected for three headform orientations of 0, 90 and 180° at three different test locations (inside, end and outside), three shock wave intensities (as indicated in the figure legend: 70, 140 and 210 kPa) and repeated four times. Adjusted R^2 *values vary from 0.82 (good fit) to 0.59 (mediocre fit) due to broad data distributions. Data are presented as average and standard deviation.*

and they do not follow the same trend as sensors mounted on the flat surface of the headform. The concave geometry of eye sockets is responsible for the compressed air entrapment and momentary stagnation during shock wave exposure, which results in extreme pressures, compared to other locations on the headform. Similar trends are also noticeable for the 90 and 180° orientations, with differences related to the geometrical factors. The amplification factors on the face exposed to the shock wave (H6, H7 for the 90°, and H5 for 180° orientation) have lower values (<3) compared to 0° orientation. The rise time for these two orientations have much

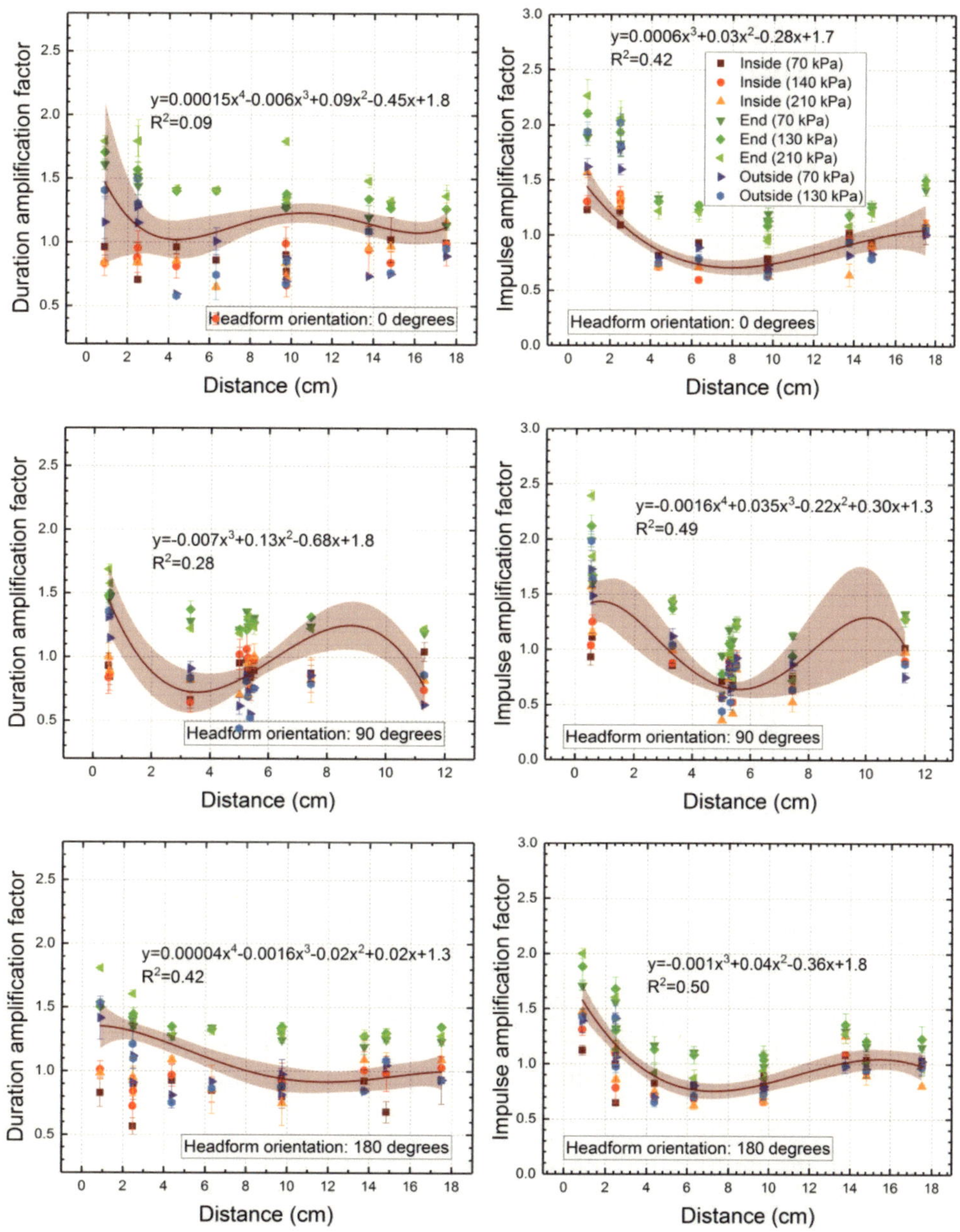

Figure 8.
Amplification factors for the duration and impulse as a function of the sensor distance on the headform. Polynomial fit performed on normalized data collected for three headform orientations of 0, 90 and 180° at three different test locations (inside, end and outside), three shock wave intensities (as indicated in the figure legend: 70, 140 and 210 kPa) and repeated 4 times. Adjusted R^2 *values vary from 0.82 (good fit) to 0.04 (bad fit) due to broad data distributions. Data are presented as average and standard deviation.*

lower values (<1 μs) and rise time values gradually increase with increased distance (**Figure 7**). Far less consistent results are obtained for the duration and impulse values (**Figure 8**). The spread of amplification factor values in all six cases is 1.0 or more, which indicates that the loading conditions at specific test location plays an important role. These variations are expressed in the goodness of fit parameters (adjusted R^2) which are presented in **Table 1**. It is clear fit parameters for peak overpressure and rise time indicate good fit, while for duration and impulse the parameters are below 0.5, which is indicative of poor-quality fit. The dispersion of the data points is the main reason, and respective data sets appear to follow stochastic distribution pattern rather than a single well-defined function.

3.3 Numerical simulation data normalization

The trends observed experimentally were confirmed using the numerical simulations. Normalized results for the spatial distribution of peak overpressures

Headform orientation	0°	90°	180°
Peak overpressure	0.80	0.61	0.75
Rise time	0.71	0.82[1]	0.79[1]
Impulse	0.42	0.49	0.50
Duration	0.09	0.28	0.42

[1]*Large spread of the data for 90 and 180° orientations.*

Table 1.
Adjusted R^2 *values for fits to different datasets presented in* ***Figures 6*** *and 7.*

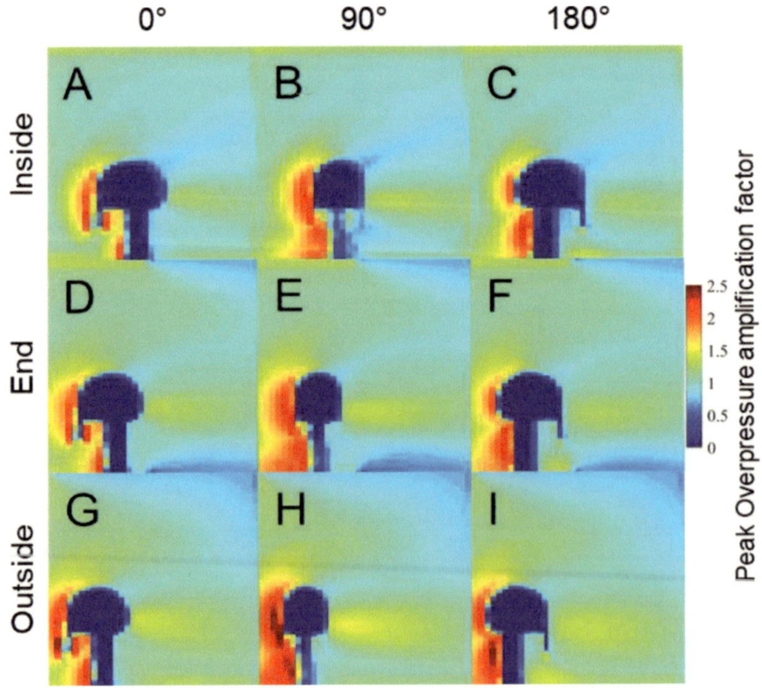

Figure 9.
Spatial distribution of the peak overpressure amplification factors for the 140 kPa exposures in the region surrounding the headform for three specimen locations and for three headform orientations.

confirmed that the highest peak overpressures are seen at the front face of the headform (**Figure 9**). Following data normalization, trends were found to be common between the three shock wave intensities, 70, 140, and 210 kPa, and, therefore, only results for 140 kPa exposures are shown. The area of stagnation is largest in the 90° orientation, followed by the 180° orientation. Additionally, for similar headform orientations, the area of stagnation increased with increasing distance from the breech, with the largest area of stagnations observed in the outside test locations and the smallest area observed at the inside location. A region of increased peak overpressures is observed behind the headform, regardless of the headform orientation or the testing location. This region indicates the area affected after the shock front wraps around the headform. Above and below that area of elevated peak overpressures, the lowest pressures are observed. This phenomenon is most apparent in the inside test location (**Figure 9A–C**), as a fan radiating behind the headform. These results corroborate the observed trends reported in **Figure 7**. The rise time was highest around the posterior face of the headform in all headform orientations and underneath the chin of the headform in the 0° orientation (**Figure 10**). In the area around the headform, the rise time was higher than that of the incident waveform. These trends were observed regardless of specimen location.

Spatial maps of the impulse amplification factors highlight the importance of testing location on specimen loading (**Figure 11**). The end location exhibited the largest impulse amplification factors and the inside location exhibited the smallest. This finding corroborates the wide spread of impulse and duration values reported in **Figure 8**. Furthermore, the impulse was highest in the 90° headform orientation, followed by the 180° orientation. The spatial distribution of the increased impulse was similar among specimen orientations and testing locations, differing in magnitudes only. A region of high impulse is seen on the front face of the headform and a region of low impulse, lower than the impulse of the incident wave, is seen behind the headform.

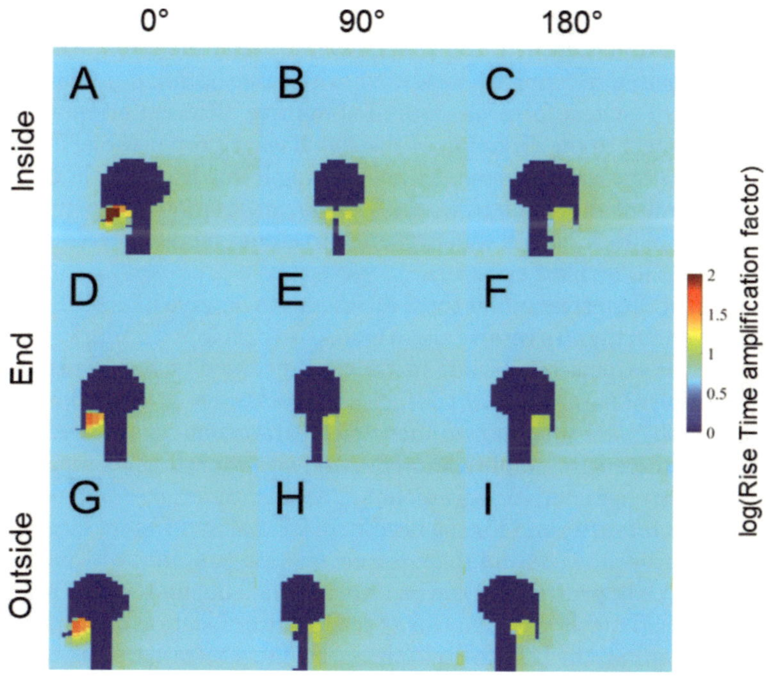

Figure 10.
Spatial distribution of the rise time amplification factors for the 140 kPa exposures in the region surrounding the headform for three specimen locations and for three headform orientations. Data are reported as the log of the rise time amplification factor, indicated the order of magnitude of the change.

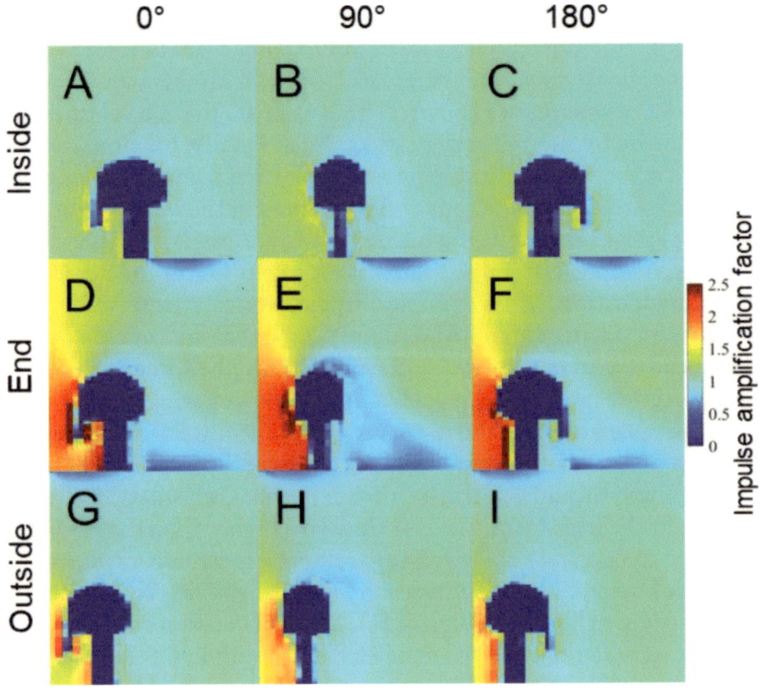

Figure 11.
Spatial distribution of the impulse amplification factors for the 140 kPa exposures in the region surrounding the headform for three specimen locations and for three headform orientations. As the duration and impulse amplification factors showed similar results, only data for the impulse are presented here.

4. Conclusions

Headform instrumented with 10 pressure sensors (mounted to measure surface pressure) was exposed to a single shock wave with three nominal intensities: 70, 140, and 210 kPa. The headform was mounted in three different orientations: 0, 90 and 180° with respect to the direction of the shock wave propagation. The effect of the headform location was evaluated by positioning it inside of the shock tube, at the end and outside of the shock tube. The headform was mounted using Hybrid III biofidelic surrogate neck, which was tightened to eliminate the inertial motion of the headform caused by blast exposure.

Comparison of the test results at three different shock wave intensities complicates data analysis even further. To resolve these issues, we developed a simple strategy of data reduction: the respective pressure parameters recorded by headform sensors are divided by equivalent parameters of the incident shock wave as defined in Eq. (1). As a result, a comprehensive set of non-dimensional parameters is generated. These non-dimensional parameters (or amplification factors) allow for direct comparison of pressure waveform characteristic parameters generated by a range of incident shock waves differing in intensity and for the headform located in different locations.

Using this approach, we found there is a correlation function which allows prediction of the peak pressure on the headform which depends only on the peak pressure of the incident shock wave (for specific sensor location on the headform), and it's independent on the headform location, and to a certain degree the orientation. Similar relationship exists also for the rise time. However, for the duration and impulse similar correlation functions do not exist.

We demonstrated via comprehensive experimental and numerical studies that three different testing locations are characterized by non-equivalent loading conditions. While it was possible to devise the transfer function for normalized peak overpressure and rise time at 0° orientation, the dispersion and clustering of data points indicates the surface pressure distribution is a function of the headform orientation where geometrical factors play important role.

Acknowledgements

This work was supported by PEO Soldier award no. W91CRB-16-C-0025.

Conflict of interest

The authors declare no conflict of interest.

Author details

Maciej Skotak, Molly T. Townsend, Eren Alay and Namas Chandra*
Department of Biomedical Engineering, Center for Injury Biomechanics, Materials, and Medicine, New Jersey Institute of Technology, Newark, NJ, USA

*Address all correspondence to: namas.chandra@njit.edu

References

[1] Vieille P. Sur les discontinuités produites par la détente brusque de gaz comprimes. Comptes Rendus de l'Académie des Sciences. 1899; **A 129**:1228-1230

[2] Celander H, Clemedson C-J, Ericsson UA, Hultman HI. The use of a compressed air operated shock tube for physiological blast research. Acta Physiologica Scandinavica. 1955;**33**(1):6-13

[3] Dryer M, Merritt DL, Aronson PM. Interaction of a plasma cloud with the Earth's magnetosphere. Journal of Geophysical Research. 1967;**72**(11):2955-2962

[4] Hooker WJ, Millikan RC. Shock-tube study of vibrational relaxation in carbon monoxide for the fundamental and first overtone. The Journal of Chemical Physics. 1963;**38**(1):214-220

[5] Kiefer J, Mizerka L, Patel M, Wei H. A shock tube investigation of major pathways in the high-temperature pyrolysis of benzene. The Journal of Physical Chemistry. 1985;**89**(10):2013-2019

[6] Park C, Appleton JP. Shock-tube measurements of soot oxidation rates. Combustion and Flame. 1973;**20**(3):369-379

[7] Kuriakose M, Skotak M, Misistia A, Kahali S, Sundaramurthy A, Chandra N. Tailoring the blast exposure conditions in the shock tube for generating pure, primary shock waves: The end plate facilitates elimination of secondary loading of the specimen. PLoS One. 2016;**11**(9):e0161597

[8] Ganpule S, Alai A, Plougonven E, Chandra N. Mechanics of blast loading on the head models in the study of traumatic brain injury using experimental and computational approaches. Biomechanics and Modeling in Mechanobiology. 2013;**12**(3):511-531

[9] Sundaramurthy A, Chandra NA. Parametric approach to shape field-relevant blast wave profiles in compressed-gas-driven shock tube. Frontiers in Neurology. 2014;**5**(253)

[10] Mejia-Alvarez R, Wilson B, Leftwich M, Martinez A, Prestridge K. Design of a fast diaphragmless shock tube driver. Shock Waves. 2015;**25**(6):635-650

[11] Tranter RS, Giri BR. A diaphragmless shock tube for high temperature kinetic studies. The Review of Scientific Instruments. 2008;**79**(9):094103

[12] Takano Y, Akamatsu T. A diaphragmless shock tube. Journal of Physics E: Scientific Instruments. 1984;**17**(8):644

[13] Swietek B, Skotak M, Chandra N, Pfister BJ. Characterization of a controlled shock wave delivered by a pneumatic table-top gas driven shock tube. Review of Scientific Instruments. 2019;**90**:075116

[14] Heufer K, Olivier H, Drumm S, Murrenhoff H, editors. A new fast acting valve for diaphragmless shock tubes. In: 28th International Symposium on Shock Waves. Manchester, UK: Springer; 2012

[15] Downey M, Cloete T, Yates A. A rapid opening sleeve valve for a diaphragmless shock tube. Shock Waves. 2011;**21**(4):315-319

[16] Gageik M, Weiss A, Klioutchnikov I, Olivier H, editors. A numerical investigation of the flow through a new fast acting valve for diaphragmless shock tubes. In: 29th International

Symposium on Shock Waves . Vol. 1. Madison, WI, USA: Springer; 2015

[17] Skotak M, Alay E, Chandra N. On the accurate determination of shock wave time-pressure profile in the experimental models of blast-induced neurotrauma. Frontiers in Neurology. 2018;**9**:52

[18] Cernak I, Savic VJ, Kotur J, Prokic V, Veljovic M, Grbovic D. Characterization of plasma magnesium concentration and oxidative stress following graded traumatic brain injury in humans. Journal of Neurotrauma. 2000;**17**(1):53-68

[19] Cernak I, Wang Z, Jiang J, Bian X, Savic J. Cognitive deficits following blast injury-induced neurotrauma: Possible involvement of nitric oxide. Brain Injury. 2001;**15**(7):593-612

[20] Kobeissy F, Mondello S, Tumer N, Toklu H, Whidden M, Kirichenko N, et al. Assessing neuro-systemic & behavioral components in the pathophysiology of blast-related brain injury. Frontiers in Neurology. 2013;**4**:186

[21] Chandra N, Sundaramurthy A, Gupta RK. Validation of laboratory animal and surrogate human models in primary blast injury studies. Military Medicine. 2017;**182**:105-113

[22] Goldstein LE, Fisher AM, Tagge CA, Zhang X-L, Velisek L, Sullivan JA, et al. Chronic traumatic encephalopathy in blast-exposed military veterans and a blast neurotrauma mouse model. Science Translational Medicine. 2012;**4**(134):134ra60-134ra60

[23] Gullotti DM, Beamer M, Panzer MB, Chia Chen Y, Patel TP, Yu A, et al. Significant head accelerations can influence immediate neurological impairments in a murine model of blast-induced traumatic brain injury. Journal of Biomechanical Engineering. 2014;**136**(9):091004-091011

[24] Alay E, Skotak M, Misistia A, Chandra N. Dynamic loads on human and animal surrogates at different test locations in compressed-gas-driven shock tubes. Shock Waves. 2017;**28**(1):51-62

[25] Sundaramurthy A, Alai A, Ganpule S, Holmberg A, Plougonven E, Chandra N. Blast-induced biomechanical loading of the rat: An experimental and anatomically accurate computational blast injury model. Journal of Neurotrauma. 2012;**29**(13):2352-2364

[26] Skotak M, Alay E, Zheng JQ, Halls V, Chandra N. Effective testing of personal protective equipment in blast loading conditions in shock tube: Comparison of three different testing locations. PLoS One. 2018;**13**(6):e0198968

[27] Needham CE, Ritzel D, Rule GT, Wiri S, Young L. Blast testing issues and TBI: Experimental models that lead to wrong conclusions. Frontiers in Neurology. 2015;**6**:72

[28] Abate G, Shyy W. Dynamic structure of confined shocks undergoing sudden expansion. Progress in Aerospace Sciences. 2002;**38**(1):23-42

[29] Arakeri J, Das D, Krothapalli A, Lourenco L. Vortex ring formation at the open end of a shock tube: A particle image velocimetry study. Physics of Fluids. 2004;**16**(4):1008-1019

[30] Moore DF, Jérusalem A, Nyein M, Noels L, Jaffee MS, Radovitzky RA. Computational biology—Modeling of primary blast effects on the central nervous system. NeuroImage. 2009;**47**:T10-T20

[31] Roberts J, Harrigan T, Ward E, Taylor T, Annett M, Merkle A. Human head–neck computational model for assessing blast injury. Journal of Biomechanics. 2012;**45**(16):2899-2906

[32] Grujicic M, Bell WC, Pandurangan B, Glomski PS. Fluid/structure interaction computational investigation of blast-wave mitigation efficacy of the advanced combat helmet. Journal of Materials Engineering and Performance. 2010;**20**(6):877-893

[33] Grujicic M, Bell WC, Pandurangan B, He T. Blast-wave impact-mitigation capability of polyurea when used as helmet suspension-pad material. Materials & Design. 2010;**31**(9):4050-4065

[34] Merkle AC, Wing ID, Roberts JC. Human surrogate head response to dynamic overpressure loading in protected and unprotected conditions. In: Herold KE, Vossoughi J, Bentley WE, editors. 26th Southern Biomedical Engineering Conference SBEC 2010, April 30–May 2, 2010, College Park, Maryland, USA. Berlin, Heidelberg: Springer; 2010. pp. 22-25

[35] Ganpule S, Gu L, Alai A, Chandra N. Role of helmet in the mechanics of shock wave propagation under blast loading conditions. Computer Methods in Biomechanics and Biomedical Engineering. 2012;**15**(11):1233-1244

[36] Nyein MK, Jason AM, Yu L, Pita CM, Joannopoulos JD, Moore DF, et al. In silico investigation of intracranial blast mitigation with relevance to military traumatic brain injury. Proceedings of the National Academy of Sciences. 2010;**107**(48):20703-20708

[37] Ott K, Voo L, Merkle A, Iwaskiw A, Wickwire A, Wester B, et al. Experimental determination of pressure wave transmission to the brain during head-neck blast tests. In: Proceedings of the ASME 2013 Summer Bioengineering Conference 26-29 June 2013. Sunriver, Oregon, USA. 2013;**55607**:V01AT10A006. DOI: 10.1115/SBC2013-14834

[38] Zhang L, Makwana R, Sharma S. Brain response to primary blast wave using validated finite element models of human head and advanced combat helmet. Frontiers in Neurology. 2013;**4**:80

[39] Przekwas A, Tan XG, Harrand V, Reeves D, Chen ZJ, S K, et al. Integrated experimental and computational framework for the development and validation of blast wave brain biomechanics and helmet protection. In: Proc HFM-207 NATO Symposium on a Survey of Blast Injury Across the Full Landscape of Military Science; Oct. 3-5, 2011; Halifax NS, Canada. 2011. pp. 34-1-34-20

[40] Grujicic M, Ramaswami S, Snipes J, Dudt P. Potential improvement in helmet blast-protection via the use of a polyurea external coating: Combined experimental/computational analyses. Proceedings of the Institution of Mechanical Engineers, Part L: Journal of Materials: Design and Applications. 2016. DOI: 10.1177/1464420716644472

[41] Ouellet S, Bir C, Bouamoul A. Direct Comparison of the Primary Blast Response of a Physical Head Model with Post-Mortem Human Subjects. QC Canada: Defence Research and Development Canada-Valcartier Research Center Quebec; 2014

[42] Ganpule S, Salzar R, Perry B, Chandra N. Role of helmets in blast mitigation: Insights from experiments on PMHS surrogate. International Journal of Experimental and Computational Biomechanics. 2016;**4**(1):13-31

[43] Zhu F, Wagner C, Dal Cengio Leonardi A, Jin X, VandeVord P, Chou C, et al. Using a gel/plastic surrogate to study the biomechanical response of the head under air shock loading: A combined experimental and

numerical investigation. Biomechanics and Modeling in Mechanobiology. 2012;**11**(3):341-353

[44] Banton R, Piehler T, Zander N, Benjamin R, Duckworth J. Comparison of numerical simulations with experiments of blast-induced pressure wave impact on a surrogate head model. In: Kimberley J, Lamberson L, Mates S, editors. Proceedings of the 2017 Annual Conference on Experimental and Applied Mechanics, Dynamic Behavior of Materials. Vol. 1. Cham: Springer International Publishing; 2018. pp. 181-187

[45] Crowley JS, Brozoski FT, Duma SM, Kennedy EA. Development of the facial and ocular countermeasures safety (FOCUS) headform. Aviation, Space, and Environmental Medicine. 2009;**80**(9):831

[46] Foster JK, Kortge JO, Wolanin MJ. Hybrid III-A biomechanically-based crash test dummy. SAE International. 1977. p. 770938. DOI: 10.4271/770938

[47] Mougeotte C, Carlucci P, Recchia S, Ji H. Novel approach to conducting blast load analyses using Abaqus/explicit-CEL. In: Center AARDE, editor. Picatinny Arsenal. NJ: Defense Technical Information Center; 2010

[48] Tan XG, Przekwas AJ, Gupta RK. Computational modeling of blast wave interaction with a human body and assessment of traumatic brain injury. Shock Waves. 2017;**27**(6):889-904

[49] Townsend MT, Alay E, Skotak M, Chandra N. Effect of tissue material properties in blast loading: Coupled experimentation and finite element simulation. Annals of Biomedical Engineering. 2018:1-14. DOI: 10.1007/s10439-018-02178-w

Chapter 8

Blast Effects on Structural Elements

María Chiquito, Anastasio P. Santos, Lina M. López and Ricardo Castedo

Abstract

Blast loads can represent a great hazard to existing structures. Their effects on structural elements can be decisive for the integrity of both the structure itself and the people within it. The behaviour of the individual elements of a building is totally different due to the heterogeneity of the materials composing them. This fact makes it necessary to carry out tests on each type of structural element in order to correctly evaluate the response of the structure. In addition, the scale effect can produce inaccurate results, making it necessary for tests to be performed on a full scale to validate the results. In this work, the results of several tests with explosives are presented, in different constructive elements, all of them carried out at full scale. These elements range from the structural elements (beams and concrete slabs) to the weak elements of a building (masonry panels).

Keywords: blast load, full scale, numerical modelling, damage assessment, structural elements

1. Introduction

In the last decades, the number of terrorist attacks made through explosive loads has increased notably. In this scenario, not only military or government facilities are targeted, but any kind of civil building may also be attacked. This fact has revealed the vulnerability of structures to explosions and blast waves and, hence, the hazard for human lives. Understanding the behaviour of structures against blast loads is absolutely essential to improve their safety.

When an explosion occurs, a sudden release of energy to the atmosphere results in a transient pressure wave or blast wave. Many structures may experience some degree of damage from air blast when the overpressure in the blast wave is about 3–5 kPa or more. The blast wave propagates outwards in all directions from the source at supersonic speed. The shock wave is characterised by an instant rise in pressure from ambient atmospheric pressure P_0 to a peak incident overpressure P_{so}. From the incident peak, the overpressure decays exponentially to the ambient value. This is the positive phase duration which is followed by a blast wind of negative pressure which sucks items back in towards the centre. This negative phase is usually much longer than the positive phase. Because negative pressures are relatively small compared with the positive phase, they are not commonly considered in blast-resistant design.

Terrorist attacks on structures may not be eliminated completely, but the effects of these attacks can be mitigated. Several techniques have been developed in recent years for this purpose, such as retrofitting or coating applications [1, 2], internal reinforcement [3, 4] or geometrical modifications in buildings [5]. It is desirable that mitigation techniques may be applied in both new and existing buildings.

The behaviour of the individual elements of a building is totally different from each other due to the heterogeneity of the materials composing them. This fact makes it necessary to carry out tests on each type of structural element in order to correctly evaluate the structure response. The behaviour of structural elements is completely different from the response of weak elements, although all of them have an influence on each other. In addition, the scale effect can produce inaccurate results, making it necessary for tests to be performed at full scale [6–8]. Furthermore, a certain repetitiveness is required to validate the results.

Due to the high costs of this kind of test, many studies in the last years have been done based on numerical modelling [finite element (FE) analysis]. However, when working in the determination of the response of structures subjected to blast loading, the complexity of the problem makes field tests mandatory to validate the results and calibrate analytical models, extending the limited scope of the field data to other design situations [9, 10].

In this work the results of tests with explosives against different constructive elements are presented, all of them carried out at full scale. These elements range from the structural elements (beams and concrete slabs) to the weak elements of a building (masonry panels). In all cases, the study carried out includes field tests, damage assessment and numerical modelling.

2. Analysis methodology

For a complete analysis of the different structural elements, the work is divided in three different and complementary stages:

- A field test campaign, always developed at full scale
- A damage assessment based on analytical and visual methods
- A numerical simulation to compare the results obtained in field tests

2.1 Test campaign

The field test campaign is designed depending on the structural element to be analysed. A previous work is carried out to calibrate particular parameters such as scaled distance, standoff distance or height of the charge. Due to the uncertainties in this kind of tests, sometimes it is necessary to correct a specific parameter along the tests to obtain the desired result. The best result to get as much information as possible from the test is to produce an intermediate damage. When a structure collapses, no information can be extracted after the explosion. The same happens when damage is minimum. Therefore, an intermediate level of damage is the best option to analyse blast effects on structures.

Elements to be tested are constructed with a representative size to avoid the scale effect. Boundary conditions are reproduced as similar as possible to a real situation. To analyse the improvement of different retrofitting materials, a non-reinforced element is always tested as a "control element". The explosive used in the tests to be reported below was PG2, a RDX-based (86%) plastic explosive for military use (similar to C4), with experimental TNT equivalent based on impulse of 0.86 [11].

Several parameters are measured in tests to characterise the explosion. Tests are monitored with pressure gauges, accelerometers, high-speed cameras, conventional cameras, laser scanner where possible or equipment to measure the velocity of detonation of the explosive with a high-speed recording device. These data are then used to calibrate the numerical simulation.

After the test, a series of measurements are collected for further analysis: the rebound value measured with a Schmidt hammer following a methodology developed by [12] combined with the surface damaged area, the spalling mass, the length of cracks and displacements.

2.2 Damage assessment

Considering all the data collected in field tests, a damage evaluation is made combining visual observations and analytical methods. The combination of both methods allows to establish a classification of damage which ranges from moderate damage to collapse. Visual observations provide information about the condition of the specimen after the explosion such as cracks, failure patterns, debris, etc. However, this kind of analysis does not allow to check if there is internal damage. To fill this gap, as analytical method, a methodology based on the rebound principle with a Schmidt hammer is used.

2.2.1 Damage based on rebound number

Methods based on the rebound principle consist of measuring the rebound of spring-driven hammer after its impact with the material surface to check its uniformity. For damage evaluation, the objective is to find a variation in the rebound number after a blast event and not a direct strength correlation [12]. For this purpose, a series of measurements on the specimen are done. One measurement location point is defined by 12 points inside a template: six for the evaluation before and six for the evaluation after the explosion (see **Figure 1** for details). Q_1 is the median of the six points evaluated before test B_i (suffix i being the corresponding number that goes from 1 to 6), and Q_2 is the median for the points after test A_i. Both medians are then compared using the Wilcoxon rank-sum test. A p-value lower than 0.05 may confirm a decrease in the rebound number after the explosion with a 95% confidence level, and hence the point analysed is damaged.

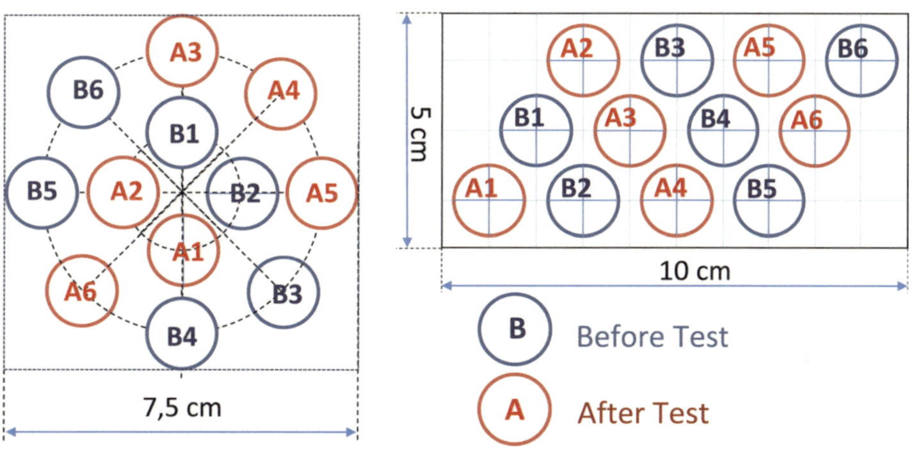

Figure 1.
Models of templates used to measure the rebound number. Left, point template used for slabs; right, point template used for bricks.

For this purpose, damage d is defined as

$$d = 1 - \frac{Q_2}{Q_1}, \text{if } p \le 0.05$$
$$d = 0, \text{if } p > 0.05 \quad (1)$$

Damage ranks from 0 to 1, 1 being a damage of 100%. In order to complete the damage analysis and based on the results obtained, a damage or contour map is created. For this purpose, a grid of 200 linearly spaced points is created for each specimen at each direction (horizontal and vertical). Then, a fitting surface to the data with a triangle-based cubic interpolation already implemented in MATLAB is calculated and plotted. With this data, a parameter called global damage (d_{200}) is obtained. This parameter is calculated with the mean of the 200 points of the grid. With this parameter, an overestimation of the mean damage is reduced.

2.2.2 Relative area of damage and crack lengths

To complete the damage evaluation, two parameters related to the surface damage can be used to rank the structure's behaviour, based on visual inspection and not in damage maps (or Schmidt hammer data). The relative damaged area (d_A) is defined as the ratio of the surface with total damage (damage equal to 1) over the total surface of the specimen. The crack length, defined as the sum of the length of all cracks presented on the specimen, can be used as another additional parameter for damage evaluation.

2.2.3 Spalling mass

After a blast event, fragmentation is the major source of wounds when the structure does not collapse. The mass of debris and the distance to which they are projected can be used as a potential damage parameter of the explosion. Therefore, the spalling fragments are classified and weighted after the tests. As this parameter is only obtained in some tests, a general comparison and conclusions with rebound data are not possible.

2.2.4 Displacements

In order to obtain more information about the structure's behaviour, tests are monitored, whenever possible, with a laser scanner measuring the geometry before and after the event. In other cases, the maximum deflection of the element, at the end of the test, is measured to assess the final degree of damage.

2.3 Numerical modelling

The destructive result of the test together with the short duration of both blast loading and response of the structure limits the opportunity for a depth understanding of the processes involved. Therefore, the use of 3D full-scale finite element modelling is fundamental to understand, predict and simulate the failure mechanism and the damage area of structural elements against blasting. Numerical models are conducted using LS-DYNA software which is based on explicit numerical methods that are suitable for solving problems associated with the high velocities and large deformations induced by blast waves.

3. Structural elements

In this section, the results of tests with explosives are presented in different constructive elements, all of them carried out at full scale. In all cases, the study includes field tests, damage assessment and numerical modelling. All tests shown here were conducted at the Technological Institute of "La Marañosa", part of the National Institute for Aerospace Technology (INTA), Spain.

3.1 Masonry walls

A total of four masonry walls tested in two different trials are included in this chapter. The walls were built inside a concrete structure, simulating the real situation of a building. For each shot two boxes or rooms (box A and box B in **Figure 2**) were built with 40 cm × 20 cm × 20 cm masonry concrete blocks joined with standard mortar. For the frontal wall, exposed to the explosion, 40 cm × 20 cm × 15 cm blocks were used. Inner dimensions of each box were 3 m × 3 m × 3 m with a 1 m separation between boxes (see **Figure 2**). Lateral and back walls were reinforced filling the hollows with mortar and steel bars (B 500S) evenly spaced in horizontal and vertical directions. The frontal wall was a continuous wall, covering the space between boxes and projecting 1 m at both sides in order to increase the impulse load over the wall. Both rooms had an access door by the central corridor. Different coatings were implemented inside the rooms for debris assessment. In each box, the inner face of the masonry wall was partially tiled (a little more than half), and the rest was pointed. After the first test, the frontal wall was rebuilt for the second trial. Explosive mass was 5 kg eq TNT and was located 0.96 m over the ground at a standoff distance of 5 m (2.92 m/kg$^{1/3}$) to the front wall.

The solution tested for each box was:

- Test M1, box A (M1A): non-reinforced wall for comparison
- Test M1, box B (M1B): aluminium mess anchored to the external face of the wall and exposed to the explosion

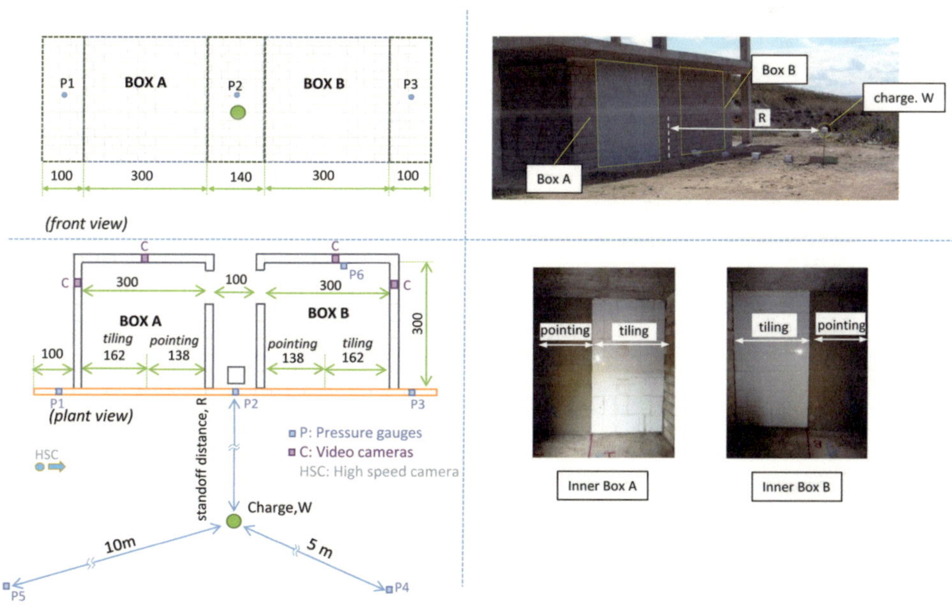

Figure 2.
Test setup and details.

- Test M2, box A (M2A): a glass fibre-reinforced sheet glued with a polyurethane adhesive in the external side of the wall
- Test M2, box B (M2B): a basalt fibre mesh attached with cementitious mortar to the inner face

3.1.1 Results and discussion

Table 1 shows the main results obtained from the test instrumentation: pressure and impulses from pressure gauges and shockwave velocity from the high-speed video.

Transducers P1, P2 and P3 were measuring the reflected pressure, while P4 and P5 were measuring the side-on or incident pressure. P6 registered the peak pressure inside box B. It can be assumed that there are not substantial pressure variations in the vertical component because the triple point height at 5 m from the charge is higher than the wall height (>3 m). The results listed on **Table 1** are from Test M1 with the exception of gauge P6 that comes from Test M2. Measurements from transducers P1 to P5 for Test M2 were lost due to a failure of the acquisition system.

Pressure results are compared with the manual from the Department of Defence, USA, UFC3-340 (DoD, 2006) [13] values and represented versus scaled distances in **Figure 3A**.

The high-speed camera was located orthogonal to the shockwave advance, at a safe distance from the explosion. This position allows to analyse the shock front propagation and its reflection in the wall face. At both tests images were recorded at 8000 frames per second with a "FastCam Sa–3", model 120 k C2, manufactured by Photron. **Figure 3B** shows the shockwave velocity evaluated compared with the values obtained from UFC 3-340.

Figure 4A shows the images of the outer face of the masonry walls after the test. Location of the boxes is marked with a discontinuous line. Damage with the rebound hammer was evaluated in 13 location points on each wall. Mean damage values obtained (d_{200}) are listed in **Table 2**. Contour maps of damage have been represented in **Figure 4B**. External damage appears at the sides, lower part in M1A and upper left corner in M1B. In test M2B damage also occurs at the left side.

Test	Location	Distance	Scaled distance	Pressure	Impulse	Shock velocity
		m	m/kg$^{1/3}$	kPa	kPa.ms	m/s
M1	P1	6.4	3.7	271	395	
M1	P2	5	2.9	386	501	
M1	P3	6.4	3.7	242	297	
M1	P4	5	2.9	149		
M1	P5	10	5.8	42	91	
M2	P6	12	7.0	5		
M1	u1	2.5	1.5			889
M1	u2	3.5	2.0			667
M2	u1	2.5	1.5			727
M2	u2	3.5	2.0			571

Table 1.
Pressure gauge results and shockwave velocity measurements.

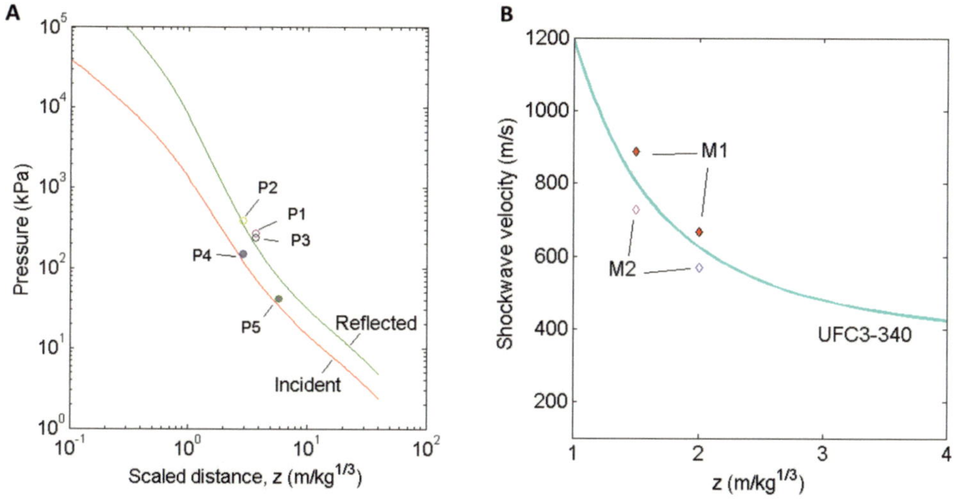

Figure 3.
(A) Pressure measurements compared with UFC3-340 data. (B) Shockwave velocity compared with UFC3-340 data.

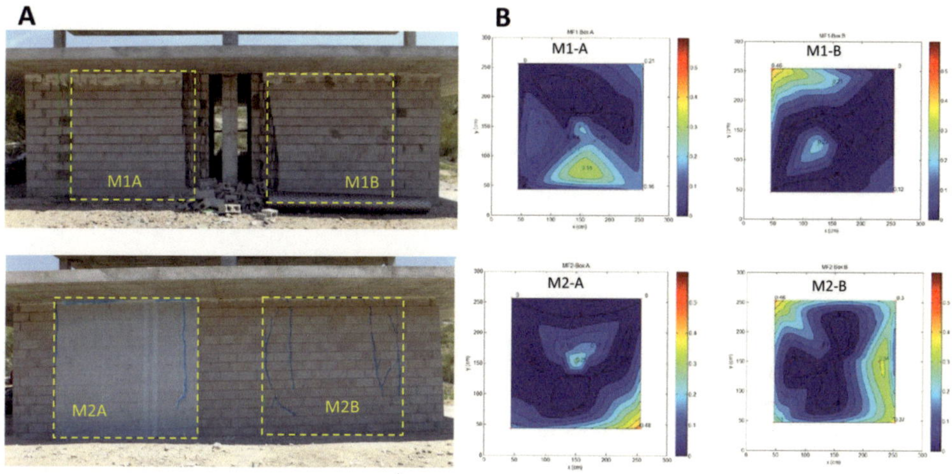

Figure 4.
(A) Results of external face of the walls after both tests. (B) Damage maps of the walls.

Test	Box	Solution	Fragment mass (kg)			Mean damage
			Total	Tiling area	Pointing area	
M1	A	No protection	63.4	56.6	6.8	0.08
M1	B	External metal mess	39.5	37.4	2.0	0.04
M2	A	Glass fibre external film	24.8	23.5	1.3	0.08
M2	B	Basalt fibre inner film	1.7	0.8	0.9	0.14

Table 2.
Debris mass and mean damage obtained with the rebound hammer.

For fragment analysis, a grid was drawn on the floor inside the boxes as shown in **Figure 5**. As can be seen, debris mainly appear in the tiling zone. Referring to the debris collected, in Test M1, it was 63.4 kg inside box A and 39.5 kg inside box B.

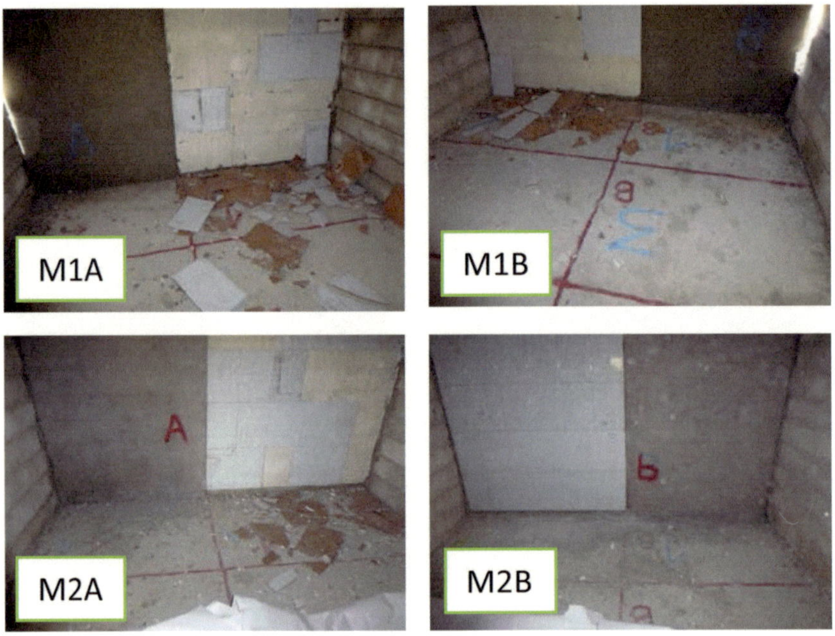

Figure 5.
Fragments inside the boxes after the test.

In Test M2, 24.8 kg of fragments were inside box A and 1.7 kg inside box B. **Table 2** shows the different fragment mass weighted at each box, specifying the mass collected inside the tiled zone and inside the pointing area. Note that in M2B there were hardly any fragments.

Comparing the results from box A with box B at M1 test, wall M1B (with the metal mess) appears with a lower mean damage (see **Figure 4B** and **Table 2**) as logic dictates because of the external protection. Furthermore, there is an important reduction in the spalling mass (from 63 kg in M1A to 40 kg in M1B).

Regarding Test M2, the influence of the external glass fibre film (M2A) and an inner basalt fibre film in wall M2B may be compared. The external film (M2A) produces a decrease on the mean damage compared with M2B, and there is a strong reduction of debris due to the inner basalt fibre film in wall M2B (from 25 kg in M2A to 1.7 kg in M2B).

Results from test one cannot be compared with those of the second test. This is due to the difference in the lapse time between the construction and test which was higher for the second. This alteration means that the results cannot be extrapolated.

3.2 Slabs and beams

Considering structural elements, different beams and slab were tested and analysed (see **Table 3** for details). The test acronym is formed by a letter that refers to the specimen to be tested as follows, S for slabs and B for beams, followed by a correlative number that reflects the trial number. Different solutions were implemented and detailed in **Table 3**. The charge mass, in terms of TNT equivalent, ranged from 2 (spherical shape) to 15 kg (cubic shape). The detonator used to initiate the explosive was located at the centre of the charge. The scaled distance went from 0.2 up to 0.79 m/kg$^{1/3}$ for slabs and beam tests.

For slabs, nine specimens of 4.40 m × 1.50 m and thickness of 0.15 m were built using a reinforced concrete with a nominal characteristic compressive strength

Test no.	Specimen	Description[a]	Standoff dist., R	Equiv. TNT, W	Scaled dist.
			m	kg	m/kg$^{1/3}$
S1, S2, S3	Slab	RC	1	2	0.79
S4	Slab	RC	1	15	0.41
S5	Slab	RC	0.5	15	0.20
S6	Slab	RC-SSA	0.5	15	0.20
S7	Slab	RC-SSG	0.5	15	0.20
S8	Slab	SFRC	0.5	15	0.20
S9	Slab	PPFRC	0.5	15	0.20
B1, B2	Beam	RC	1	2	0.79
B3	Beam	RC	1	4	0.63
B4	Beam	RC	1	15	0.41
B5	Beam	RC	0.5	15	0.20

[a]RC, reinforced concrete; SSA, steel sheet anchored on top; SSG, steel sheet glued on top; SFRC, steel fibre-reinforced concrete; PPFRC, polypropylene fibre-reinforced concrete

Table 3.
Characteristics of the tests.

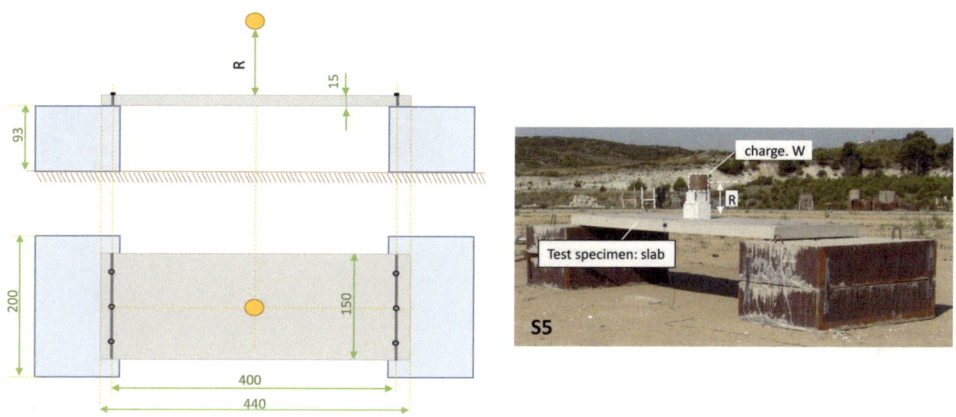

Figure 6.
Slab test setup.

of 25 MPa. Slabs were anchored at the edges to concrete blocks pretending a rigid support. **Figure 6** shows the test layout. The reinforced meshing was constructed with a 12 mm diameter rebar spaced 150 mm × 150 mm. In addition, some slabs had an extra reinforcement for protection purposes. In slabs S6 and S7, a steel sheet of 1.50 m × 1.50 m and a thickness of 10 mm were located on the surface directly exposed to the blast. Fibre-reinforced concrete was used in slabs S8 and S9. For S8 steel fibres were used and polypropylene fibres for S9. The main characteristics of concrete, steel and fibres used are detailed in **Table 4**.

The explosive charge for standoff distances of 1 m was located hanging from a rope, while for lower distances, the charge was placed on an expanded polystyrene cubic support. Note that the expanded polystyrene cube used for tests S4–S9 is just acting as a placeholder and the possibility of a shockwave reflection on the cube can be rejected.

In case of beams, five reinforced concrete specimens were tested. **Figure 7** shows the test setup and the charge placement. Beams had a section of 0.25 m × 0.20 m

	Concrete	Steel bars	Steel sheet	SFRC	PPFRC
Density (kg/m^3)	2300	7850	7850	—	—
Compressive strength (MPa)	25	—	—	44.16	43.33
Tensile strength (MPa)	3.7	550	550	8.12	5.62
Young modulus (MPa)	273 × 10^2	200 × 10^3	200 × 10^3	—	—
Yield strength (MPa)	—	500	275	—	—
Tangent modulus (MPa)	—	425	1850	—	—
Fibres dosage in concrete (kg/m^3)	—	—	—	120[a]	9[a]
Length (mm)	—	—	—	50±1[a]	48±1[a]
Diameter (mm)	—	—	—	1 ± 0.1[a]	0.84 ± 0.1[a]
Specific volume fraction (%)	—	—	—	1.5[a]	1[a]

[a]*Fibre properties*

Table 4.
List of materials' characteristics used in slabs and beams.

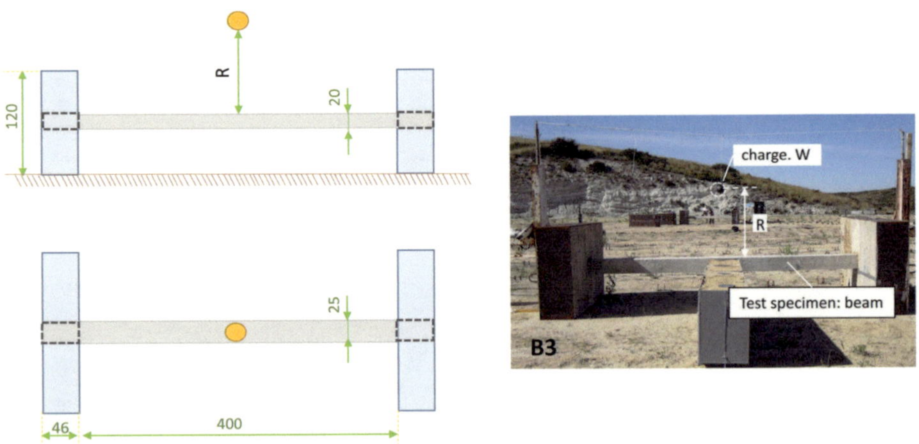

Figure 7.
Beam test setup.

with spans of 4.00 m. The reinforcement consists of four holding bars of 12 mm in diameter and 6 mm diameter vertical stirrups spaced 150 mm along. The materials used here (concrete and steel) were the same than those used in the slabs (see **Table 4**). Each beam was fixed at both ends to a concrete block. For beams B1, B2 and B3, the explosive was hanged from a rope; however, for trial B4, the explosive was situated on an expanded polystyrene cubic support.

In order to assure the equipment integrity, measuring devices were only used in tests with low explosive charge (S1–S3 and B1, B2). To calibrate and validate numerical models, these data are necessary. Three accelerometers were used in each specimen and two pressure transducers with ablative protection. In these tests, high-speed camera resulted as not very useful as the "fire ball" covered all the screen, but information about the blast phenomena may be always extracted from videos.

3.2.1 Results and discussion

Table 5 shows the different parameters of damage calculated. A rank is also included for the assessment result, based on authors' experience and others [3, 8, 10, 14, 15], and considering d_{200} and d_A parameters, no damage, moderate or severe.

Test no.	Scaled distance	Assessment result	Points	Damage		Area damaged	
	(m/kg$^{1/3}$)		n	d_n	d_{200}	$d_{A,top}$	$d_{A,bottom}$
S1, S2, S3	0.79	No damage	19	0.00	0.00	0	0
S4	0.41	Moderate	19	0.23	0.10	0.03	0.07
S5	0.20	Severe	19	0.38	0.27	0.07	0.27
S6	0.20	Severe	19	0.53	0.30	0.22	0.32
S7	0.20	Severe	19	0.42	0.25	0.11	0.28
S8	0.20	Severe	19	0.40	0.27	0.05	0.13
S9	0.20	Severe	19	0.48	0.28	0.06	0.15
B1	0.79	No damage	9	0.00	0.00	0	0
B2	0.79	No damage	9	0.01	0.00	0	0
B3	0.63	No damage	9	0.01	0.00	0	0
B4	0.41	Moderate	9	0.11	0.08	0.08	—
B5	0.20	Severe	9	0.37	0.22	0.20	—

Table 5.
Damage parameters for slabs and beam tests.

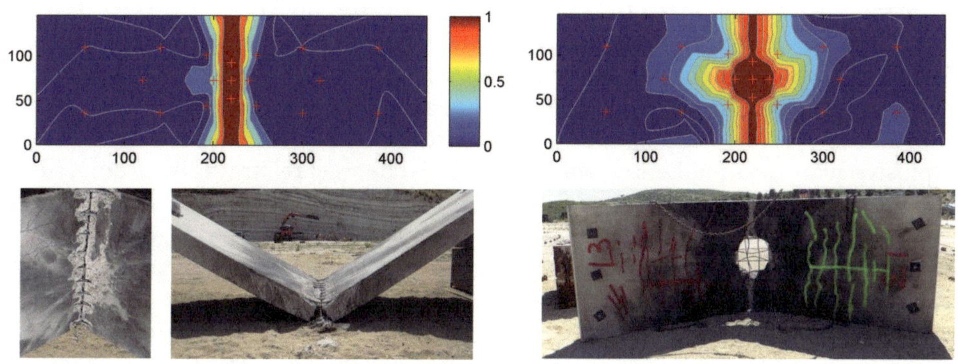

Figure 8.
Results for S4 and S8 tests and the comparison with damage maps. Z being the scaled distance.

While one of the two parameters is below 0.05, the damage is classified as no damage; no damage may denote appearance of little cracks, while concrete remains almost intact. If one of the parameters is above 0.05 and below 0.15, the damage is considered as moderate referring to specimens with important affected area and moderate to heavy concrete spalling. At last, if one parameter is above 0.15, the damage is severe considered as full spalling of the concrete thickness. Damage d_n has been evaluated at 19 points on the slab top surface. For slabs S1–S3, results were no damage. Only some little cracks appeared at the bottom face of slab S3. **Figure 8** shows the explosion effects on two slabs tested and the output from the damage maps.

As this methodology reflects, when reducing the scaled distance, damage d_{200} increases for slabs S1 to S5. These are the slabs without additional reinforcement. For slabs with steel sheet (S6 and S7), there is an increase of the area damaged in about 15 and 4%, respectively, when compared with the same configuration without steel sheet, S5. Nevertheless, the bottom part results defer up to 5%. At the same scaled distance, steel and polypropylene fibre slabs (S8 and S9), the evaluated damage d_{200} is almost the same compared with slab S5. Furthermore, when

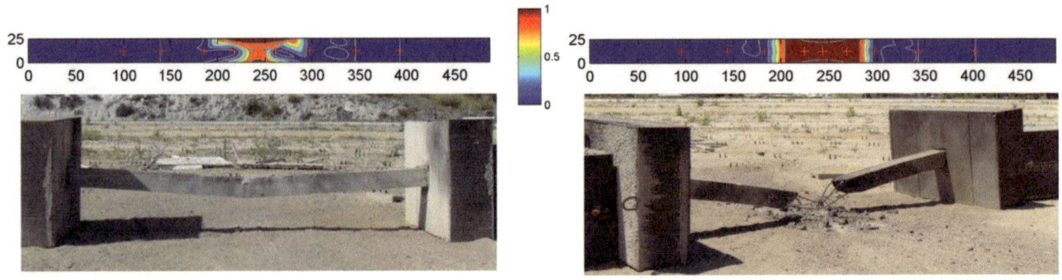

Figure 9.
Results for B4 and B5 tests and the comparison with damage maps. Z being the scaled distance.

considering the area of damage on the top surface, the values are also very similar. However, at the bottom parts, the differences are noticeable.

Table 5 also shows the data obtained after blast events over five beams tested. In this case, no additional reinforcement material was implemented in beams, so the study is only based on reinforced concrete (RC) beams. Damage d_n has been evaluated at 9 points on the top surface of the beam, which was the one exposed to the explosion. As in the slab test results, when the scaled distance decreases, all the damage parameters evaluated increase. Tests B1–B3 resulted in no damage after the evaluation. **Figure 9** shows the damage maps with some photos of the results for tests B4 and B5, with scaled distances of 0.41 and 0.20 m/kg$^{1/3}$, respectively. In these tests, the only area damaged (d_A) measured was the top surface (the surface directly exposed to the blast) due to the arrangement of the test, obtaining similar results as those with slabs for higher scaled distances.

In the numerical simulation, reinforced concrete slabs and beams were reproduced using 3D solid Lagrangian elements (using LS-DYNA software), while the explosive event is implemented with CONWEP. The steel rebar was incorporated into the concrete mesh using the LS-DYNA command: *Constrained_Lagrange_in_Solid. Using this feature, we assume a perfect bond, i.e., no sliding, due to short event time and high blast pressures. To save computational time, and taking advantage of the symmetry conditions, only one half along the major axis of each slab is considered. More details about the model can be found in Castedo et al. [15]. The explosive charge has been introduced into the model by the *Load_Blast_Enhanced (LBE) function. This function is based on the Kingery-Bulmash empirical blast data [16]. The plasticity-based material model CSCM was used to simulate the concrete material in the slabs. The CSCM model has the option of inputting your own material properties or requesting default material properties for normal strength concrete introducing only the compressive strength f_c (**Table 4**). In this case, the second option was used. In this material type, the erosion is activated when its internal damage (plastic strain in LS-PrePost) reaches 99%. The steel bars were modelled using the piecewise linear plasticity material. This material model deals with the stress strain by a bilinear curve considering strain rate effects.

Results of numerical modelling can be used to compare the results of field tests, and if good agreement appears, results can be extrapolated to other cases without the need for verification by testing. As can be seen in **Figure 10**, models reproduce well the result of the field test. Furthermore, data extracted about area of damage are acceptable when compared with those measured in the field with errors up to 15% [17].

For beam modelling, considering the existing symmetry planes in the structure, and in order to reduce as much as possible the computational time, only a quarter of beam is created. It has been verified that the results for the models of a quarter with two symmetries, a half with a symmetry and the complete model give similar results.

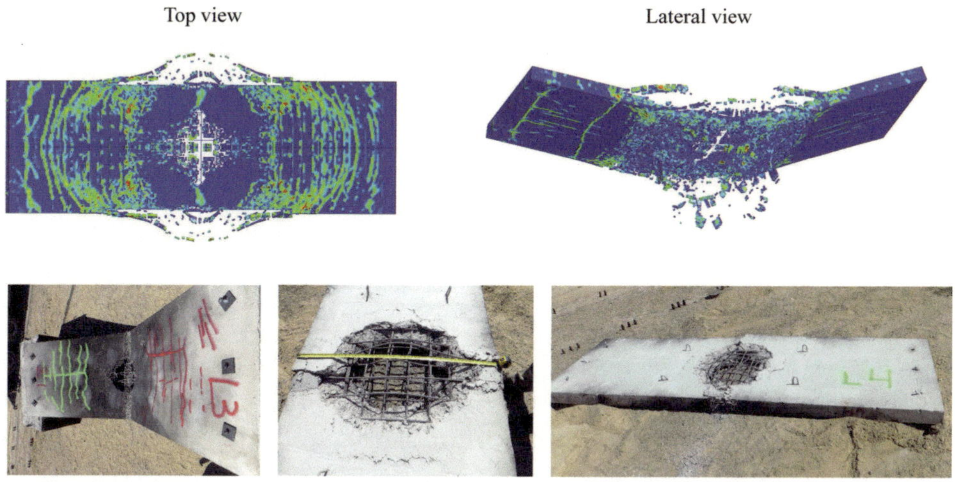

Figure 10.
Numerical modelling of S8 showing plastic strain and the comparison with the photographs of the test.

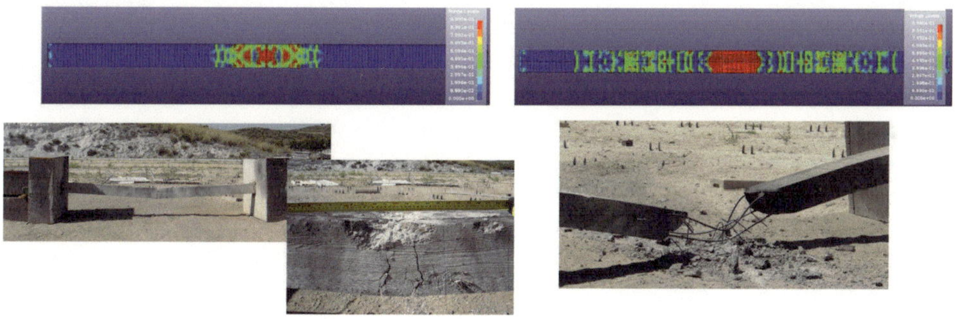

Figure 11.
Comparison of the numerical modelling for tests B4 and B5 and the test results.

The explosive charge has been introduced into the model in the same way that for slabs. Also, the same material models have been used for concrete and steel. **Figure 11** shows the plastic strain produced by the numerical modelling for beams B4 and B5 and the comparison with the photographs after the tests. As happened with the damage assessment and contour maps, results are in good agreement with those shown and measured in field tests.

3.3 Structural integrity

Once different structural elements and weak elements of a building have been tested, it would be desirable to know the behaviour of a complete structure subjected to a blast load. This kind of test is really difficult to implement due to the element dimension but mostly because of the high cost. In the test shown here, the behaviour of a frame structure after the removal by blasting of one of its columns is analysed experimentally, and the correspondence of this scenario to the theoretical calculation situation is checked. The structure analysed was a two-storey reinforced concrete structure. Each slab had dimensions of 14.35 m long by 11.00 m wide including this measure 2 metres of cantilever of the slab on each side of the structure. Each slab is supported by six reinforced concrete columns with a square section of 35 cm × 35 cm. The whole structure is built on site with reinforced concrete

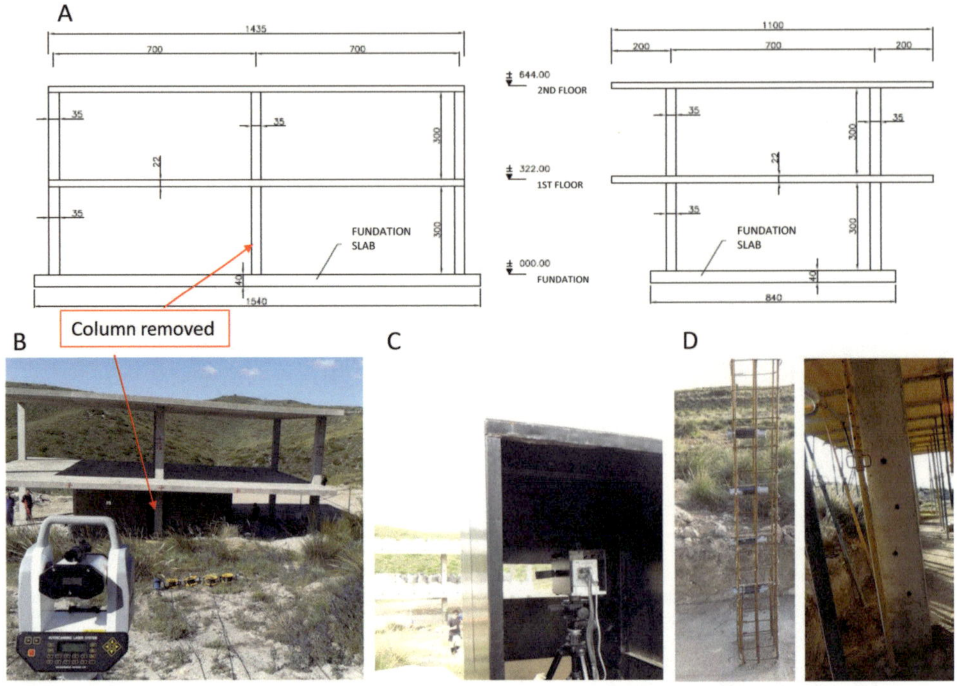

Figure 12.
(A) Geometrical details of structure. (B) Laser profiler used to measure the permanent strains. (C) High-speed camera. (D) Details of the column to be removed.

with a nominal characteristic compressive strength of 25 MPa and rebar with a yield strength of 500 MPa. **Figure 12** shows the constructive details.

The column to be instantly removed with explosives was one of the central columns of the main floor (see **Figure 12**). For the demolition of the column, four PG2 cylindrical loads of 200 g each were used. Insensitive instantaneous electric detonators were used and connected in series.

The test was monitored with the following equipment:

- Laser profiler: robotic total station that allowed to scan the structure and to evaluate the permanent strains.
- Accelerometers: four accelerometers connected to a fast acquisition system.
- Seismographs: recording of particle velocity (vibration) and air wave pressure (microphone). Three triaxial and one uniaxial geophones.
- High-speed cameras: analysis of the different phases of movement and evaluation of displacement and velocity. Two high-speed cameras were used.

3.3.1 Results and discussion

After the column was removed, the structure collapsed. The accelerometers and seismographs were trapped in the debris and could not record anything. No measure was possible with the laser profiler either, as there was no permanent strain after the collapse to be measured. The high-speed camera video made possible to evaluate the falling speed of the slab placed on the column removed. This measurement contributed to the calibration of the corresponding simulations.

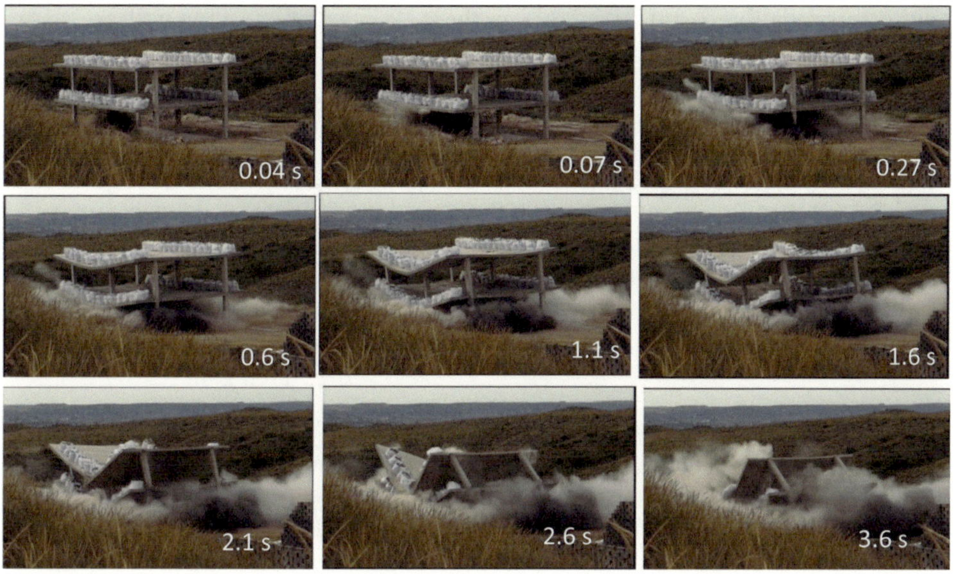

Figure 13.
Sequence of the test recorded with a conventional camera.

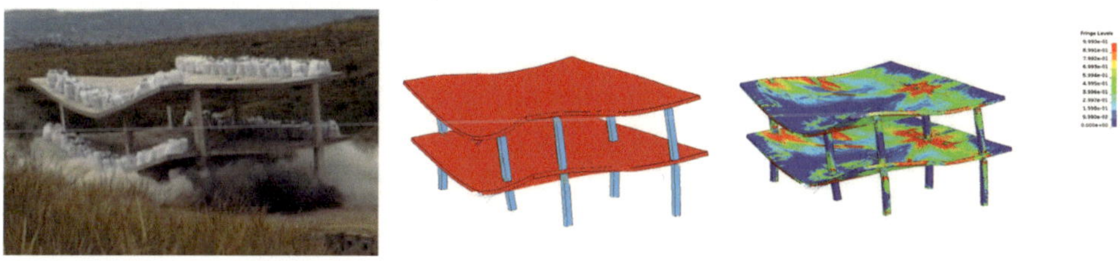

Figure 14.
Photograph of the structure collapse vs. 3D numerical model of the structure at time 1.1 seconds.

The test was also recorded with two conventional cameras. **Figure 13** shows the sequence of the test with one of the cameras and the moment at which structure collapsed.

As the result of the test was the total collapse of the structure, no more information could be extracted apart from the visual data (videos recorded), hence the importance of achieving the intermediate level of damage in the trials. To reproduce and analyse the behaviour of the structure, numerical modelling was used. As in the previous cases of this work, a 3D model with Lagrangian mesh and CONWEP blast load description is developed. Also, the same material models (CSCM for concrete and piecewise linear plasticity material model for rebar) have been used. More details about the model are included in Bermejo et al. [18]. The models used for the reinforced concrete structure have proven their capacity to reproduce the complete process of the progressive collapse. **Figure 14** shows a frame from the video recorded at 1.1 seconds and the comparison with the numerical model at the same instant.

4. Conclusions

Several works with different constructive elements have been reported here. The complexity of the phenomena studied and the magnitude of the consequences highlight the great importance of the research in this field. For this purpose, only numerical modelling is not acceptable, as there is a need of checking with real cases. Scale factor is another big problem, making it necessary to develop the tests at full scale. Furthermore, a good design of tests is essential to achieve an intermediate level of damage and, hence, to extract as much information as possible.

A total of 16 different tests have been carried out at full scale with different constructive elements. Finally, one test over a complete structure of a building has been developed and analysed. In most cases, numerical modelling and damage assessment have been made with the data extracted from the tests. In addition, numerical modelling of these tests can be used as a tool for predicting possible scenarios and analyse threats with similar conditions to those tested here.

Acknowledgements

This research has been conducted under the SEGTRANS project funded by the Centre for the Development of Industrial Technology (CDTI), an agency of the Spanish government. We would like to thank TAPUSA, EUROESTUDIOS, KV Consultores and Sacyr, which contributed to the design and execution of the tests. We also would like to thank all the staff at La Marañosa (ITM-INTA) for their help at the testing site.

Author details

María Chiquito*, Anastasio P. Santos, Lina M. López and Ricardo Castedo
E.T.S.I. de Minas y Energía, Universidad Politécnica de Madrid, Madrid, Spain

*Address all correspondence to: maria.chiquito@upm.es

References

[1] Buchan PA, Chen JF. Blast resistance of FRP composites and polymer strengthened concrete and masonry structures – a state-of-the-art review. Composites Part B: Engineering. 2007;**38**(5):509-522

[2] Silva PF, Lu B. Improving the blast resistance capacity of RC slabs with innovative composite materials. Composites Part B: Engineering. 2007;**38**(5):523-534

[3] Tabatabaei ZS, Volz JS, Baird J, Gliha BP, Keener DI. Experimental and numerical analyses of long carbon fiber reinforced concrete panels exposed to blast loading. International Journal of Impact Engineering. 2013;**57**:70-80

[4] Pantelides CP, Garfield TT, Richins WD, Larson TK, Blakeley JE. Reinforced concrete and fiber reinforced concrete panels subjected to blast detonations and post-blast static tests. Engineering Structures. 2014;**76**:24-33

[5] Hajek R, Foglar M. Numerical and experimental analysis of the effect of rigid barriers on blast wave propagation. Journal of Structural Engineering. 2015;**141**:1-12

[6] Schenker A, Anteby I, Gal E, Kivity Y, Nizri E, Sadot O. Full-scale field tests of concrete slabs subjected to blast loads. International Journal of Impact Engineering. 2008;**35**(3):184-198

[7] López LM, Segarra P, Castedo R, Sanchidrián JA, Navarro J, Chiquito M. Performance of different constructive solutions against explosions in full-scale masonry walls. In: Proceedings of the 11th International Symposium on Rock Fragmentation by Blasting. Australia: Sydney; 2015

[8] Chiquito M, López LM, Castedo R, Santos AP, Pérez-Caldentey A. Blast effects and damage characterization on reinforced masonry walls at full-scale. WIT Transactions on The Built Environment. 2018;**180**:113-123

[9] Thiagarajan G, Kadambi AV, Robert S, Johnson CF. Experimental and finite element analysis of doubly reinforced concrete slabs subjected to blast loads. International Journal of Impact Engineering. 2015;**75**:162-173

[10] Wang W, Zhang D, Lu F, Wang SC, Tang F. Experimental study and numerical simulation of the damage mode of a square reinforced concrete slab under close-in explosion. Engineering Failure Analysis. 2013;**27**:41-51

[11] López LM, Sanchidrán JA, Piedra LJ, Rios J. Shock wave pressure in underground explosions. Explosives and blasting technique. In: Holmberg R, editor. Proceedings of the EFEE 2nd World Conference. Prague: Taylor & Francis; 2003. pp. 441-447

[12] López LM, Castedo R, Chiquito M, Segarra P, Sanchidrián JA, Santos AP, et al. Post-blast non-destructive damage assessment on full-scale structural elements. Journal of Nondestructive Evaluation. 2019;**38**(1):30. DOI: 10.1007/s10921-019-0572-y

[13] Department of the Army. Structures to Resist the Effects of Accidental Explosions, UFC 3-340-02. Washington, DC: U.S. Department of the Army, Navy and Air Force; 2008

[14] Zhao CF, Chen JY. Damage mechanism and mode of square reinforced concrete slab subjected to blast loading. Theoretical and Applied Fracture Mechanics. 2013;**63**:54-62

[15] Castedo R, Segarra P, Alañon A, Lopez LM, Santos AP, Sanchidrian JA. Air blast resistance of full-scale slabs with different compositions: Numerical

modeling and field validation. International Journal of Impact Engineering. 2015;**86**:145-156. DOI: 10.1016/j.ijimpeng.2015.08.004

[16] Kingery CN, Bulmash G. Air Blast Parameters from TNT Spherical Air Burst and Hemispherical Surface Burst. Maryland: U.S. Army Ballistic Research Laboratory; 1984

[17] Castedo R, Lopez LM, Chiquito M, Santos AP. Full-scale reinforced concrete slabs: Blast effects, damage characterization and numerical modelling. International Journal of Safety and Security Engineering. 2019;**9**(1):50-60. DOI: 10.2495/10.2495/SAFE-V9-N1-50-60

[18] Bermejo M, Santos AP, Goicolea JM. Development of practical finite element models for collapse of reinforced concrete structures and experimental validation. Shock and Vibration. 2017;**2017**:9. DOI: 10.1155/2017/4636381